Industrial Map of
China's Energy

Industrial Map of
China's Energy

China Industrial Map Editorial Committee
China Economic Monitoring & Analysis Center

W **World Scientific**

NEW JERSEY · LONDON · SINGAPORE · BEIJING · SHANGHAI · HONG KONG · TAIPEI · CHENNAI

Published by

World Scientific Publishing Co. Pte. Ltd.

5 Toh Tuck Link, Singapore 596224

USA office: 27 Warren Street, Suite 401-402, Hackensack, NJ 07601

UK office: 57 Shelton Street, Covent Garden, London WC2H 9HE

British Library Cataloguing-in-Publication Data

A catalogue record for this book is available from the British Library.

中国能源产业地图 2010–2011

Originally published in Chinese by Social Sciences Academic Press

Copyright © Social Sciences Academic Press, 2011

INDUSTRIAL MAP OF CHINA'S ENERGY

ISBN 978-981-4425-35-3

In-house Editor: Dong Lixi

Typeset by Stallion Press

Email: enquiries@stallionpress.com

Printed in Singapore by B & Jo Enterprise Pte Ltd

China Industrial Map Editorial Committee

Chairperson:

Jiang Yao   Chairman, Shanghai Instrumentation & Electronics Group Holding Company

Committee Members:

Shen Laiyun   Spokesman, National Bureau of Statistics of China (NBS)

Director, Department of Comprehensive Statistics, NBS

Zhang Zhongliang   Director, Department of Finance and Construction, NBS

Zuo Xiaolei   Chief Economist, China Galaxy Securities

Xu Tiefu   Director, China Economic Monitoring & Analysis Center, NBS

Pan Jiancheng   Deputy Director, China Economic Monitoring & Analysis Center, NBS

Secretary General:

Yu Jiangang   Vice President, Shanghai Instrumentation & Electronics Group Holding Company

China Economic Monitoring & Analysis Center

The China Economic Monitoring & Analysis Center of the National Bureau of Statistics (NBS) is the authority on research and publication of economic indexes in China. The Center's primary functions include monitoring the status of national economic prosperity, forecasting future development trends, identifying new trends and issues within economic sectors, as well as publishing related economic prosperity figures. They publish *China Monthly Economic Indicators* (CMEI), a report which provides economic statistics on the all facets of society and guides the monitoring and analysis of the status of economic prosperity throughout China.

Editor's Note

Energy is the material base human beings rely on to produce and lead their lives which, together with materials and information, is regarded as one of the three main elements for modern social development. The energy consumption level is a significant indicator to measure a nation's economic state, scientific and technological progress, and people's living standards. As for China's current economy, the energy industry has stood out, during the past ten year's heavy industrialization, with a status equal to the lifeline of the national economy. Seen from this perspective, the energy industry is a subject of core status for the research on the complicated trend of economic operation in China.

2010 is the final year of the Eleventh Five-Year Plan, and also the year of preparation for the coming economic restructuring. On the one hand, driven by the execution of the economic stimulus package and various policies and measures against the global economic crisis and for maintaining steady and rapid economic development, China's economy has bounced back from the impact of the worldwide economic crisis during the years 2008 and 2009, and witnessed a sustained and rapid growth. Impelled by the macro economy, energy economics has recovered in a positive direction and continues to be further strengthened: a rapid growth in energy production, new highs in output, a continuous recovery of energy consumption, and a universal rise in energy prices. To summarize, the fundamental plane of the energy industry has been witnessing a promising prospect ever since 2007.

On the other side of the coin, however, the global quantitative easing brings in not only a rapid increase in energy demand, but a return of the looming shadow of inflation. Inflation ensuing from the boom has become the main target of state regulation. Moreover, since the energy industry itself lies in the upper reach of the industrial chain, its price control has become the main target in preventing inflation. Meanwhile, we have noticed that after the government took initiatives to regulate the economy, and put forth a series of measures on energy conservation and reducing pollutant discharge, the energy consumption per unit of output value has some-what fallen. This is the only road for economic restructuring. Against this general background, this publication, a map of China's energy industry, demonstrates to the readership what this industry has undergone, together with necessary interpretation.

This publication reserves the compiling style of the Rongtian Information industrial map series: with tables and charts as the main illustration means, combined with concise words; a comprehensive, dense and detailed reflection of the status quo, showing the features and trends of development of China's energy industry. In addition, compared with last year, the structure of this publication has certain innovations as follows: it describes the energy industry from two perspectives, looking at the overview and the individual industries; it is well larded with the state of the global industry, attempting to give readers a clearer and more elaborate map of China's energy industry during the years 2009 and 2010.

This book is comprised of two major parts: for the overview part, it compares and analyzes the major statistics of the energy industry worldwide and that in China from six perspectives, namely, resources, production, consumption, trade, economy and environment; for the part concerning specific industries, it follows the trends of industries and demonstrates the specific industrial situations of petroleum, natural gas, coal and new energy from different angles such as production, consumption, imports/exports and typical enterprises of the industry.

A well-known investor once mentioned that God relies only on statistics and what behind them. Therefore, based on the detailed and factual statistical analysis, it is our greatest honor to represent, in a direct, visual and concise way, to readers the latest status of the energy industry in China.

Contents

Overview

Chapter 1   Energy Resources

Section 1: The world's and China's major energy resources

Distribution of world energy resource reserves

In 2009, the remaining proven oil reserves were world over 181.73 billion tons, increasing by 6.4% compared with that in the previous year. In terms of the current exploitation level, the exploitable lifetime of the proven reserves is 45.7 years; in 2010, the remaining proven oil reserves the world over were 188.79 billion tons, an increase of 3.9% compared with that in the previous year, and the exploitable lifetime is 58.6 years.

In 2009, the proven exploitable coal reserves the world over were 826.00 billion tons, maintaining the same level as that in the previous year, with an exploitable lifetime of 119 years; in 2010, the proven exploitable coal reserves the world over were 860.94 billion tons, increasing by 4.2% compared with that in 2009, and the exploitable lifetime is 118 years.

In 2009, the remaining proven natural gas reserves the world over the were 186.6 trillion m³, increasing by 1.2% compared with that in the previous year, and the exploitable lifetime of the proven reserves is 62.8 years; in 2010, the remaining proven reserves of

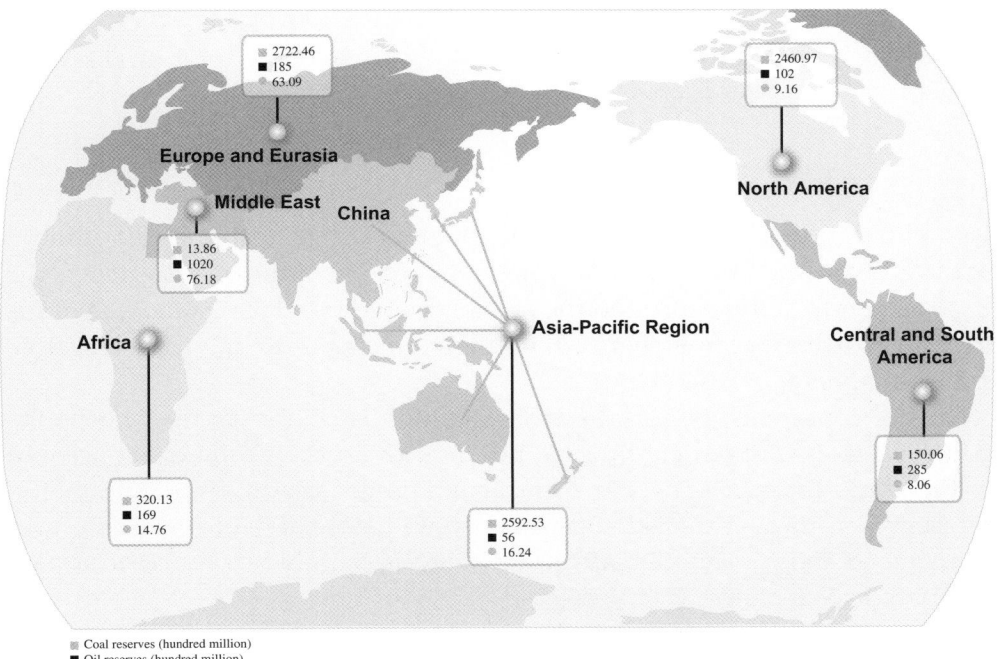

Distribution of remaining proven reserves of energy in various regions the world over by the end of 2009

Coal reserves (hundred million)
Oil reserves (hundred million)
Natural gas reserves (trillion cubic meters)

3

Distribution of remaining proven reserves of energy in various regions the world over by the end of 2010

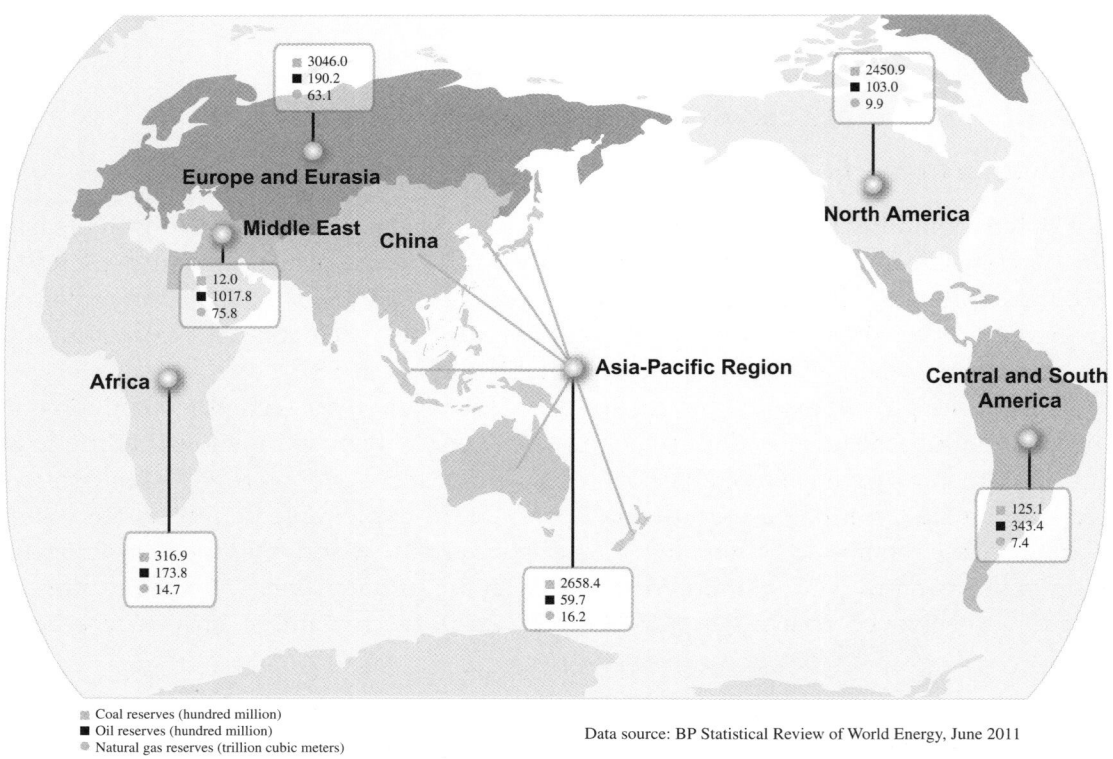

Coal reserves (hundred million)
Oil reserves (hundred million)
Natural gas reserves (trillion cubic meters)

Data source: BP Statistical Review of World Energy, June 2011

were natural gas were 187.1 trillion m³, with a slight rise of 0.3%, and the exploitable life-time of the proven reserves is 58.6 years.

Oil

On the whole, the distribution of world oil resources is in extreme disequilibrium; in terms of the Eastern and Western Hemispheres, about 3/4 of the oil resources are concentrated in the Eastern Hemisphere and the rest in the Western Hemisphere; in terms of the Southern and Northern Hemispheres, the oil resources are mainly concentrated in the Northern Hemisphere.

In terms of continents, oil resources are mainly distributed in the Middle East connecting Europe, Asia and Africa. In 2009, the total reserves in this region accounted for 56.1% of the world total. Besides that, Central and South America is a region with a relatively rapid growth in world crude oil reserves, with the share of remaining proven oil reserves rising to 15.7%. The increase in its share is mainly attributed to a large-scale reserve growth in Venezuela.

In 2010, the oil reserves in Middle East still ranked first, accounting for over half of the world total, though with a slight drop in its share.

Distribution of remaining proven oil reserves in various regions the world over

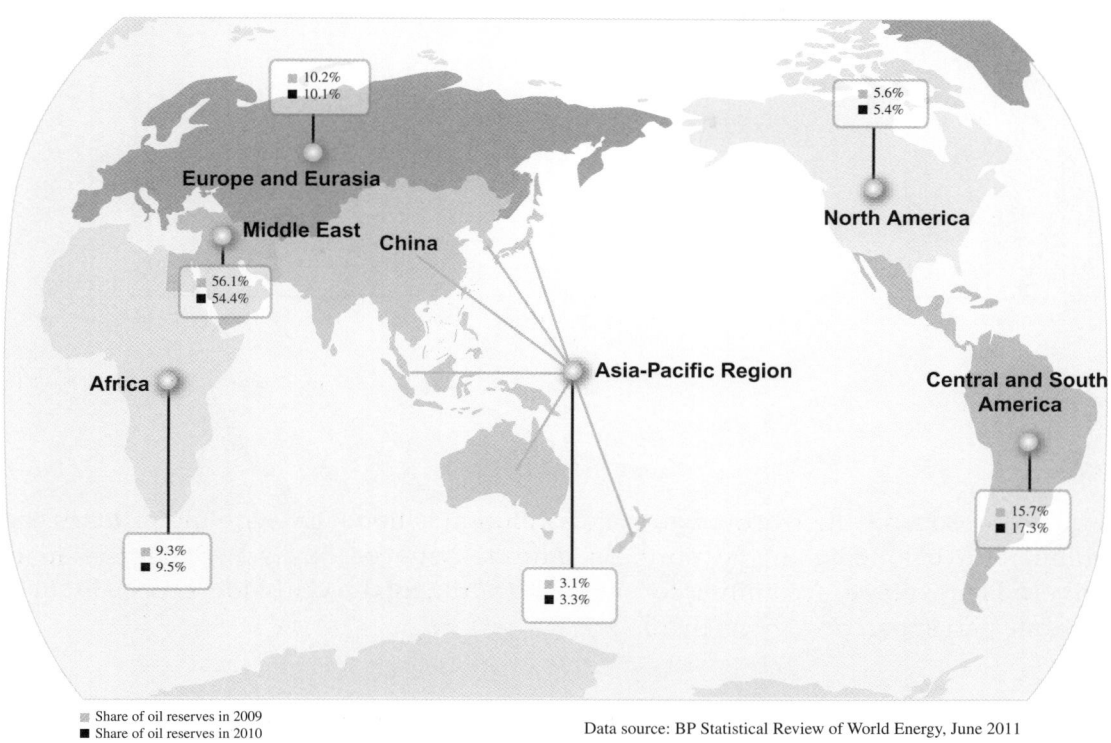

■ Share of oil reserves in 2009
■ Share of oil reserves in 2010

Data source: BP Statistical Review of World Energy, June 2011

In 2009, among the top 10 in world crude oil reserves, five countries were from the Middle East, namely, Saudi Arabia, Iran, Iraq, Kuwait and the United Arab Emirates (UAE). In 2010, the oil reserves in the Middle East remained in the front ranks of the world.

Comparison of oil reserves in the top 10 countries and China in 2009

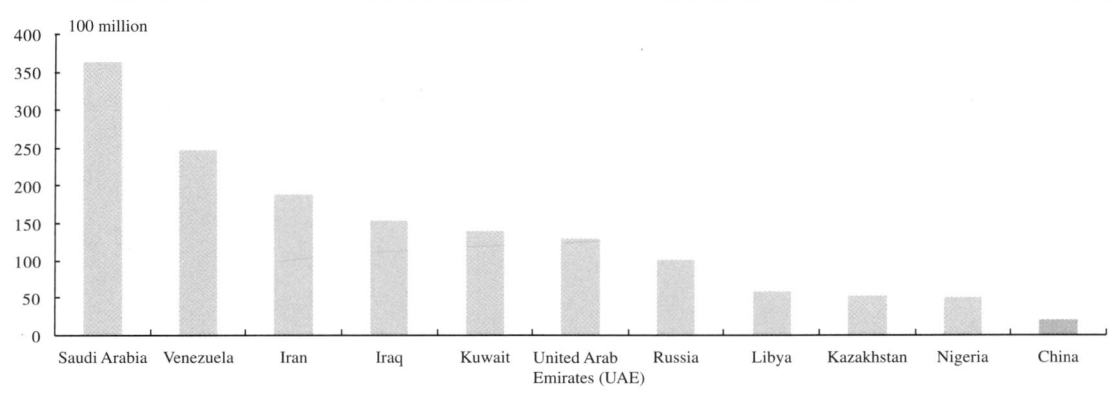

Data source: BP Statistical Review of World Energy, June 2010

Comparison of oil reserves in the top 10 countries and China in 2010

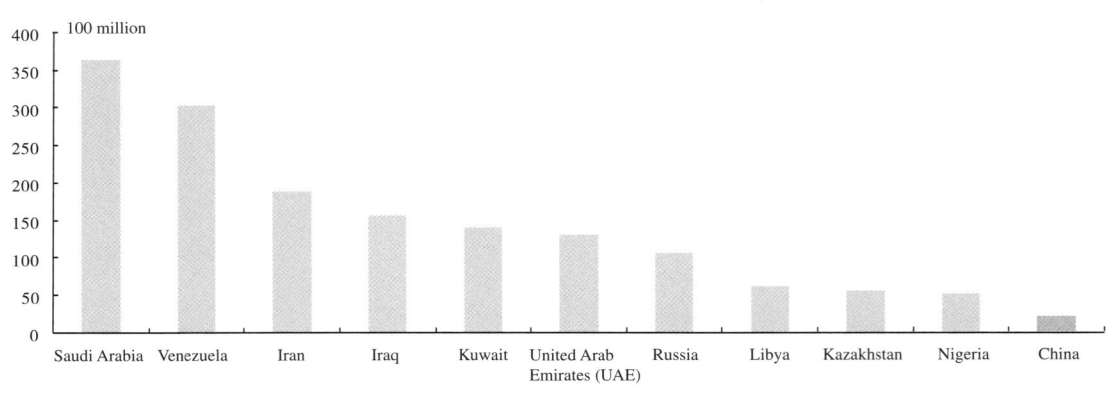

Data source: BP Statistical Review of World Energy, June 2011

Coal

As for coal resources the world over, the distribution is not even, with the primary coal-accumulation belt lying in the northern latitude between 30 and 70 degrees in the Northern Hemisphere, accounting for over 70% of the total worldwide, especially in the temperate and subarctic regions there.

Distribution of remaining proven coal reserves in various regions the world over

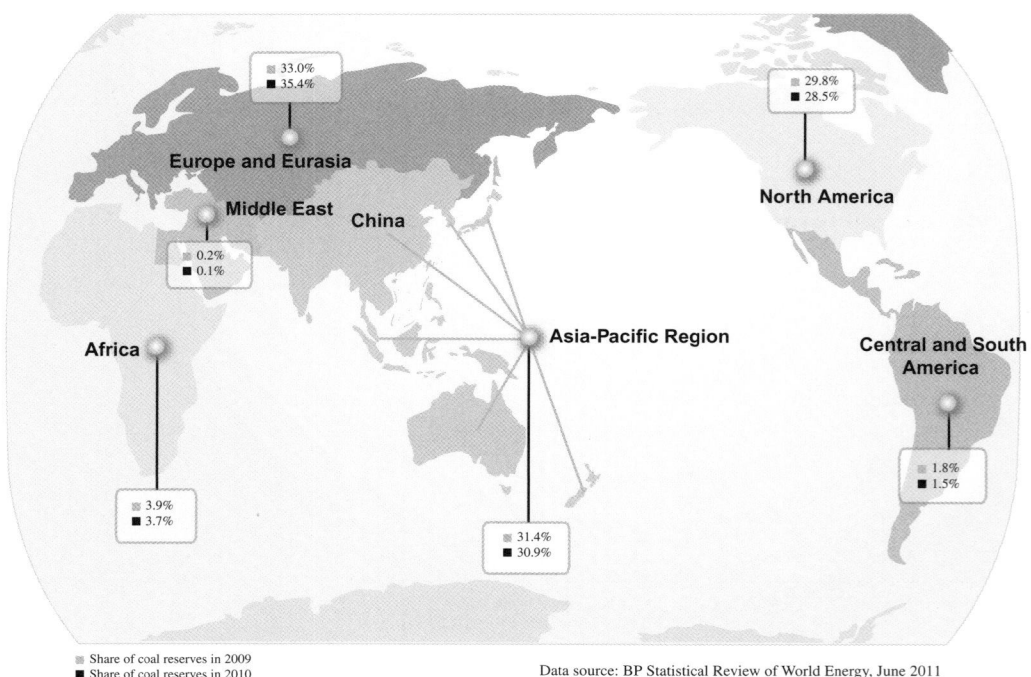

Share of coal reserves in 2009
Share of coal reserves in 2010

Data source: BP Statistical Review of World Energy, June 2011

A comparison between these continents shows that coal resources in the three continents in the Northern Hemisphere are relatively abundant: in 2009 and 2010, in terms of the proven coal reserves, the Asia-Pacific Region accounted for 31.4% and 30.9% respectively, that of North America 29.8% and 28.5% respectively, and that of Europe and Eurasia 33.0% and 35.4% respectively. In contrast, coal resources in the continents of the Southern Hemisphere are smaller in amount: Central and South America had a share of 1.8% and 1.5% respectively, with the sum of coal resources in Africa and the Middle East accounting for 4.1% and 3.8% of the world total respectively.

In terms of possession of coal resources, America, Russia and China are the most abundant in coal resources.

In 2009, America boasted the largest amount of proven coal reserves, about 238.31 billion tons, accounting for 28.9% of the world total. Following was Russia with its proven coal reserves of 157.01 billion tons, accounting for 19% of the total worldwide; the next was China with its proven coal reserves of 114.50 billion tons, accounting for 13.9% of the total.

In 2010, these three countries still ranked among the world's top three in coal reserves, with their proven reserves basically unchanged.

Comparison of proven reserves of world top 10 coal-producing countries in 2009

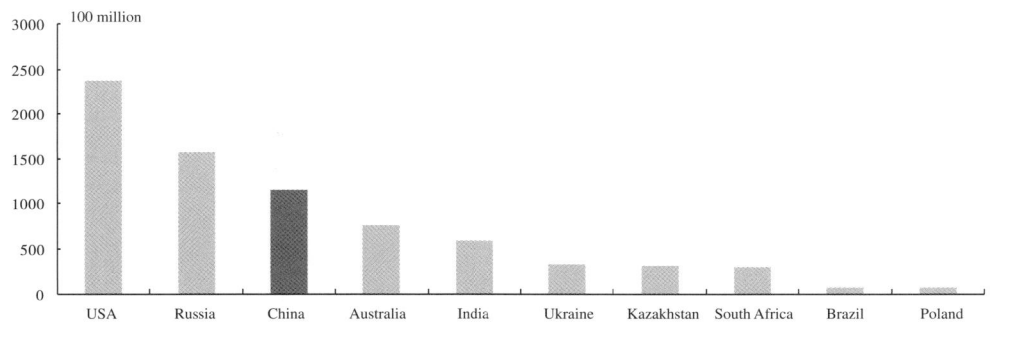

Data source: BP Statistical Review of World Energy, June 2010

Comparison of proven reserves of world top 10 coal-producing countries in 2010

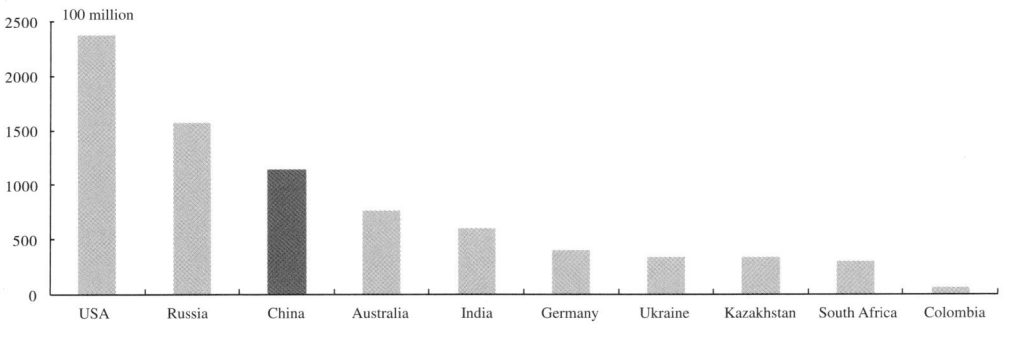

Data source: BP Statistical Review of World Energy, June 2011

Natural gas

In terms of reserves in these continents, the Middle East and Europe and Eurasia enjoy the most abundant natural gas in the world.

In 2009, proven reserves of natural gas in the Middle East were 75.7 trillion m³, accounting for 40.6% of the world total, and that were Europe and Eurasia 63.0 trillion m³, with a share of 33.8%.

In 2010, the proven reserves of natural gas in these two regions witnessed a small-scale increase, accounting for 40.5% and 33.7% of the world total respectively.

Distribution of remaining proven reserves of natural gas in various regions the world over

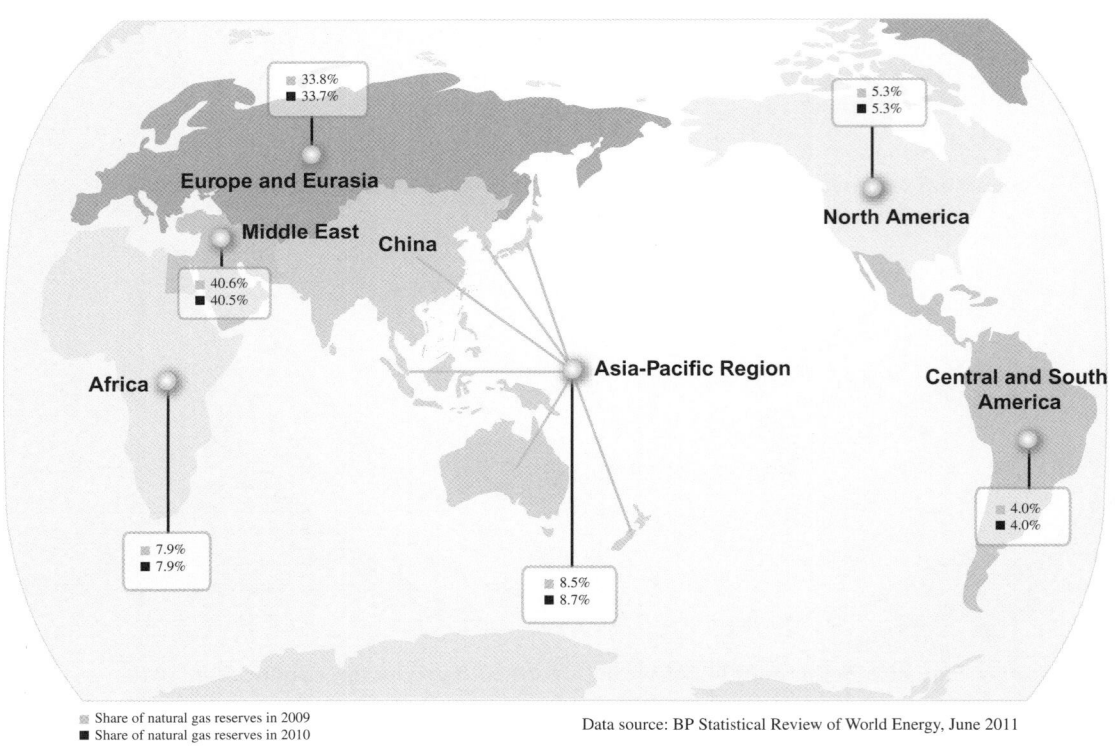

Share of natural gas reserves in 2009
Share of natural gas reserves in 2010

Data source: BP Statistical Review of World Energy, June 2011

The world's top three host countries of natural gas resources are Russia, Iran and Qatar.

In 2009, the proven natural gas reserves of these three countries were 44.4 trillion m³, 29.6 trillion m³ and 25.37 trillion m³, with their share of the world total being 23.7%, 15.8% and 13.5% respectively.

In 2010, the three countries still ranked as the world's top three in proven natural gas reserves. Among them, Russia's natural gas proven reserves rose to 44.8 trillion m³, while Iran and Qatar remained at the same level as in the previous year.

Comparison of natural gas reserves in the top 10 countries and China in 2009

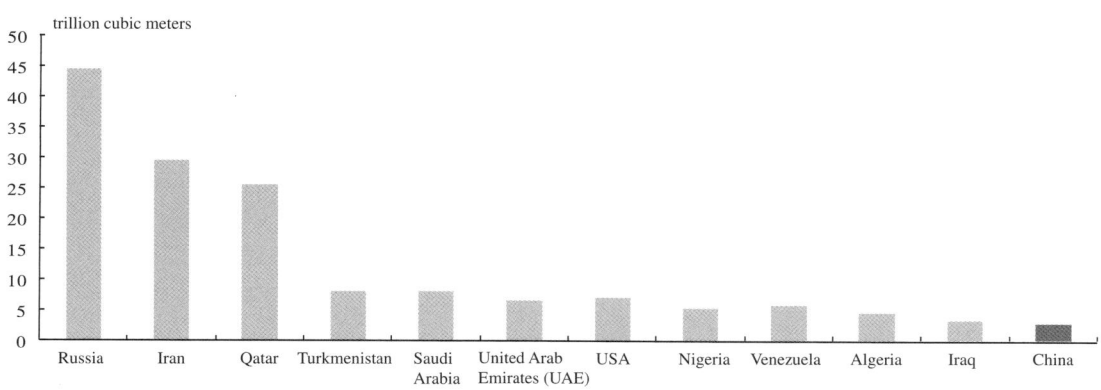

Data source: BP Statistical Review of World Energy, June 2010

Comparison of natural gas reserves in the top 10 countries and China in 2010

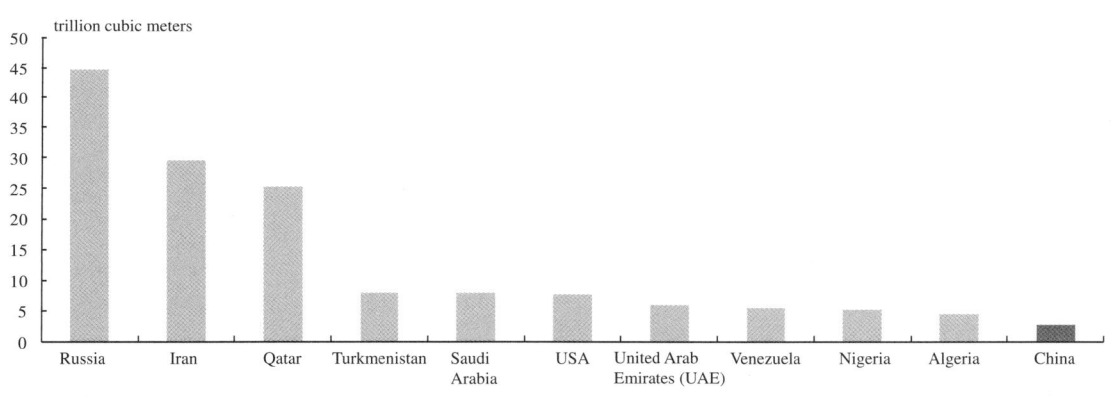

Data source: BP Statistical Review of World Energy, June 2011

Distribution of energy resources in China

Coal

China abounds in coal resources, with diverse varieties and widespread distribution. All the 32 provinces/regions/municipalities except Shanghai have their share of coal resources. However, the distribution of coal resources in China is rather uneven, with a profile of "rich in the north and west while scarce in the south and east", mostly concentrated in Shanxi Province and the northwest territories.

In North China, the majority of coal resources is distributed all over the whole or part of six provinces and regions — Inner Mongolia, Shanxi, Shaanxi, Ningxia, Gansu and Henan, their sum of reserves accounting for about 50% of the total nationwide. In South China, coal resources are mainly distributed in Guizhou, Yunnan and Sichuan, which enjoy a share of 90% of the total coal resources in South China, and whose proven reserves in sum account for over 90% of the total proven reserves in South China.

Major distribution of coal resources in China

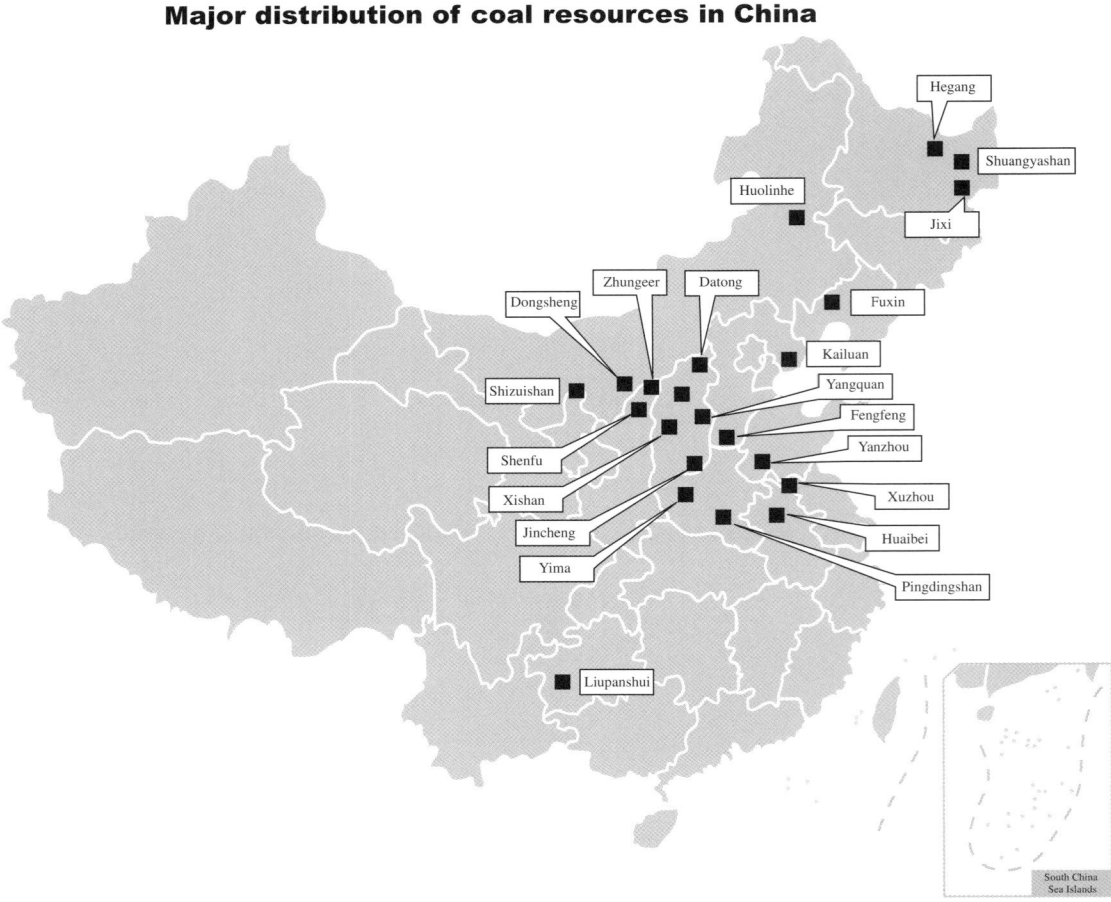

During the period of the Eleventh Five-Year Plan, our country launched key construction projects founding 13 large-scale coal bases in eastern Inner Mongolia, Shendong, northern Shaanxi, western Shandong, Henan, Yunnan-Guizhou, and the two-Huai area, etc. In 2010, the total output of the 13 coal bases amounted to 2.8 billion tons, accounting for about 87.5% of the total coal output all over the country.

During the execution of the Twelfth Five-Year Plan, our country will promote, in an orderly way, the construction of 14 large-scale coal bases, among which Xinjiang will transform from a coal reserve base to the 14th large-scale base officially. It is our objective to try our best so the output of these bases in all will obtain a 90% share of the total output nationwide.

Distribution of 13 large-scale coal bases in China

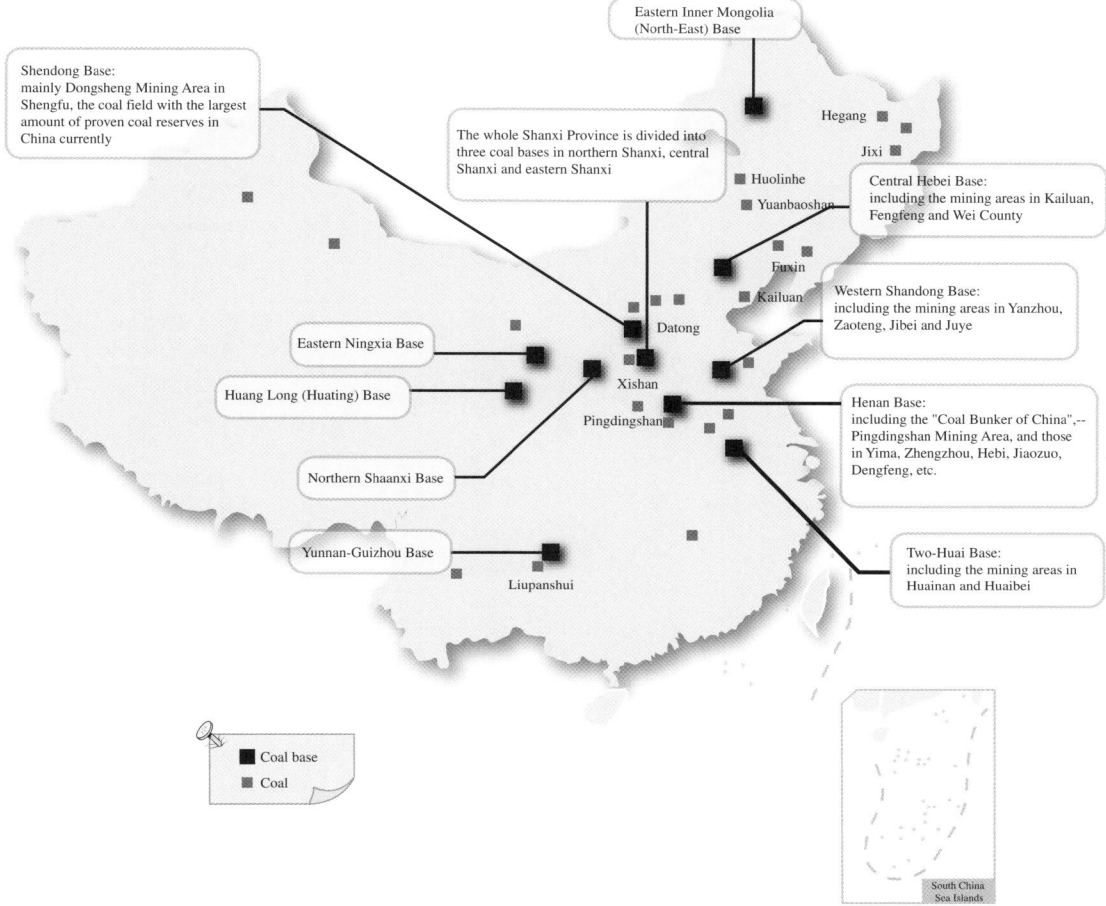

Eastern Inner Mongolia (North-East) Base

Shendong Base: mainly Dongsheng Mining Area in Shengfu, the coal field with the largest amount of proven coal reserves in China currently

The whole Shanxi Province is divided into three coal bases in northern Shanxi, central Shanxi and eastern Shanxi

Hegang

Jixi

Huolinhe

Yuanbaoshan

Central Hebei Base: including the mining areas in Kailuan, Fengfeng and Wei County

Fuxin

Kailuan

Western Shandong Base: including the mining areas in Yanzhou, Zaoteng, Jibei and Juye

Eastern Ningxia Base

Datong

Huang Long (Huating) Base

Xishan

Henan Base: including the "Coal Bunker of China",-- Pingdingshan Mining Area, and those in Yima, Zhengzhou, Hebi, Jiaozuo, Dengfeng, etc.

Pingdingshan

Northern Shaanxi Base

Yunnan-Guizhou Base

Two-Huai Base: including the mining areas in Huainan and Huaibei

Liupanshui

Coal base

Coal

South China Sea Islands

Oil and gas

The distribution of China's oil and gas resources is unbalanced, with the present large and medium-sized oil-gas fields basically in areas north of the Changjiang River, mainly the northern part, such as North China, West China and Northeast China. Their total output accounts for over 90% of the national total output. As for the eastern part of the areas north of the Changjiang River, it is rich in oil while scarce in gas. For the western regions, the distribution of oil and gas is balanced.

The oil resources are mainly distributed over five large-scale basins, i.e. the Bohai Sea Gulf Basin, Songliao Basin, Tarim Basin, Ordos Basin and Zhungeer Basin, whose total output of exploitable oil resources accounts for 73% of the total the nation over. By the end of 2009, China's remaining proven reserves of oil were 2 billion tons, with a reserve-production ratio of 10.7; by the end of 2010, the former figure remained the same, also 2 billion tons, with a reserve-production ratio of 9.9.

Distribution of China's oil and gas resources

By the end of 2009, China's remaining proven reserves of natural gas were 2.5 trillion m³, with a reserve-production ratio of 28.8; by the end of 2010, the former figure changed to 2.8 trillion m³, with a reserve-production ratio of 29.0.

The natural gas resources are more concentrated than oil resources. For the former, they are mainly distributed over three big basins, namely, the Tarim Basin, Sichuan Basin and Ordos Basin. The sum of the exploitable oil resources in these three basins boasts a share of over 55% of the total nationwide.

Section 2: Renewable energy resources in China

Renewable energy includes hydroenergy, biomass energy, wind energy, solar energy, geo-thermal energy and ocean energy, boasting great resource potential, less environmental pollution and sustainable use, all of which are highly important energy resources beneficial to the harmonious development between humanity and nature. China ranks the first worldwide in exploitable installed capacity and annual power generation. In addition, it also abounds in renewable energy resources like solar energy, wind energy and biomass energy.

During the years from 1978 to 2010, in China's energy consumption structure, fossil energy consumption was central; coal consumption accounted for over 68% of total primary energy consumption, with the total share of oil and natural gas above 22%, and that of new energy and renewable energy less than 9%.

In 2009, the Chinese government made a solemn commitment to the international community that it would spare no effort to the share of non-fossil energy in primary energy consumption increase to around 15% by 2020.

Hydroenergy

China boasts numerous rivers, with abundant runoff and significant drop, making it rich in hydroenergy. It ranks first the world over both in hydroenergy deposits and exploitable hydroenergy reserves.

In China, among the most abundant in hydroenergy are the southwest areas of Sichuan, Yunnan, Tibet and Guizhou, whose total hydroenergy accounts for 66.7% of the nationwide total exploitable hydroenergy; next followed by are the south-central areas of Hubei, Guangxi Zhuang Autonomous Region, Guangdong Province, etc., with a total share of 12%; the northwest areas of Qinghai, Xinjiang Uygur Autonomous Region, Gansu and Shaanxi, etc., whose total share is 9.5%.

China is planning 13 large-scale hydropower bases in terms of the river basins of hydroenergy, i.e. Jinshajiang River, Yalong River, Dadu River, Wu River, upper reach of the Changjiang River, Qingjiang River, Hongshuihe River, Langcang River, upper reach of the Yellow River, middle reach of the Yellow River, Hunan, Fujian-Zhejiang-Jiangxi river basins, northwest river basins and Nujiang River.

Share of hydroenergy of the major river basins in China

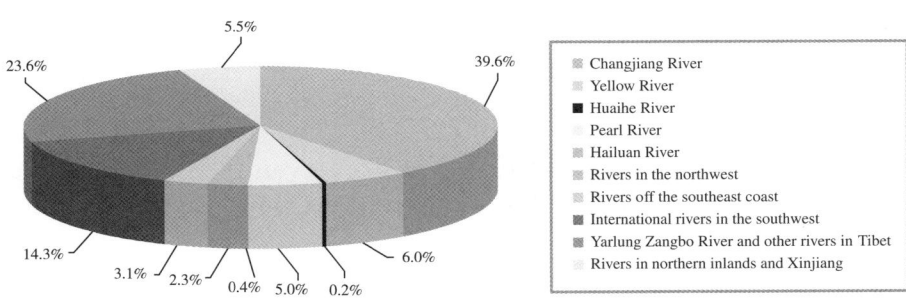

Data source：Lifeblood--50 Years' Hydraulics of People's Republic of China, Three Gorges.

Distribution of large-scale hydropower bases in China

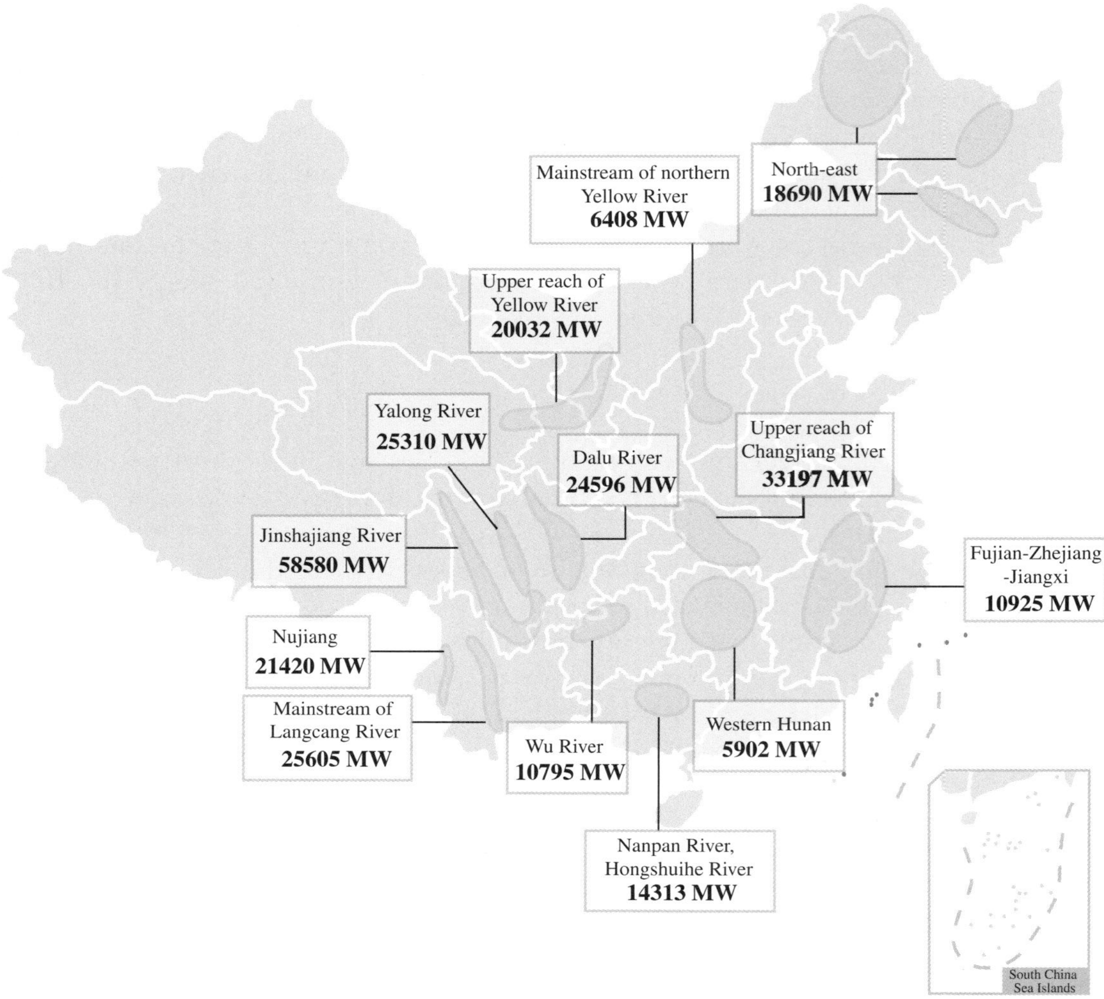

Mainstream of northern
Yellow River
6408 MW

North-east
18690 MW

Upper reach of
Yellow River
20032 MW

Yalong River
25310 MW

Dalu River
24596 MW

Upper reach of
Changjiang River
33197 MW

Jinshajiang River
58580 MW

Fujian-Zhejiang
-Jiangxi
10925 MW

Nujiang
21420 MW

Mainstream of
Langcang River
25605 MW

Wu River
10795 MW

Western Hunan
5902 MW

Nanpan River,
Hongshuihe River
14313 MW

South China
Sea Islands

When the People's Republic of China was founded, the total exploited installed capacity of hydropower was only 360,000 kilowatts (kW). In September 2004, this total surpassed the 100 million kW mark, thus the country with the largest amount of hydropower making China.

The scale of hydropower construction in the Eleventh Five-Year period was unprecedented. A batch of large-sized hydropower stations were constructed one after another, such as Longtan, Jianghong, Goupitan, Laxiwa, Xiaowan and Pubugou. The number of newly added units that have been put into operation during these five years approximates to the sum total of that in the first 95 years since China began to exploit hydropower. Besides, the construction of large-sized and oversized hydropower stations in Xiangjiaba and Jinxing has been successively launched.

Trend of China's total installed capacity of hydropower

10,000 kW

- Installed capacity of hydropower generation equipment
- Growth margin

Values (bars): 7,935 (2000); 8,301 (2001); 8,607 (2002); 9,490 (2003); 10,524 (2004); 11,742 (2005); 13,029 (2006); 14,823 (2007); 17,260 (2008); 19,629 (2009); 21,340 (2010)

Growth margin: 4.60% (2001); 3.70% (2002); 10.20% (2003); 10.90% (2004); 11.60% (2005); 11.00% (2006); 13.80% (2007); 16.40% (2008); 13.70% (2009); 8.70% (2010)

Data source: National Bureau of Statistics of China (NBS)

Wind energy

Abundant belt in the three-north region (Northeast, North and Northwest China): The power density of wind energy is above 200–300 W/m², with a utilizable hourage of over 5000 h. The formation of this abundant belt is mainly attributed to its geographic position in the middle and high latitudes.

Abundant belt off the coast and the surrounding islands: The annual power density of available wind energy is over 200 W/m², with a utilizable hourage of around 7000–8000 h. Benefiting from the funneling effect of the Taiwan Strait, the southeast coast and the surrounding islands form an optimal abundant belt of wind energy. Our country possesses a coastline of over 18,000 km, and more than 6000 islands, and has a promising prospect of exploring and utilizing wind energy.

Inland abundant belt: The power density of wind energy normally is below 100 W/m², with a utilizable hourage of less than 3000 h.

The wind energy market in China has maintained growth momentum. In 2009, its cumulative installed capacity of wind-powered electricity generation was

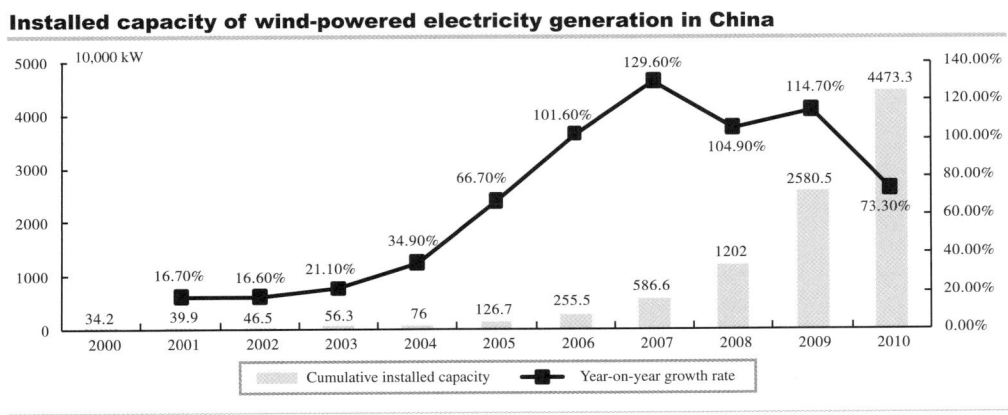

Installed capacity of wind-powered electricity generation in China

10,000 kW

- Cumulative installed capacity
- Year-on-year growth rate

Cumulative installed capacity (bars): 34.2 (2000); 39.9 (2001); 46.5 (2002); 56.3 (2003); 76 (2004); 126.7 (2005); 255.5 (2006); 586.6 (2007); 1202 (2008); 2580.5 (2009); 4473.3 (2010)

Year-on-year growth rate: 16.70% (2001); 16.60% (2002); 21.10% (2003); 34.90% (2004); 66.70% (2005); 101.60% (2006); 129.60% (2007); 104.90% (2008); 114.70% (2009); 73.30% (2010)

Data source: NBS

25.805 million kW, increasing by 114% compared with the same period the previous year, thus making it the fourth consecutive year of doubling installed capacity. In 2010, this cumulative installed capacity continued to grow, up to 44.733 million kW, a 73.3% rise compared with the same period the previous year.

Solar energy

China lies in the east of Eurasia in the Northern Hemisphere, mainly located within the temperate and subtropical zones, and abounds fairly in solar energy. According to the long-term cumulative data from over 700 domestic meteorological stations, the annual total solar radiation all over the country is around 3.35×10^3 to $8.40 \times 10^3 \, MJ/m^2$, with an average of about $5.86 \times 10^3 \, MJ/m^2$.

The isoline starts from northeast Inner Mongolia to the western foothills of the Greater Khingan Mountains, crossing southward to the northwest side of Beijing, then westward in the south to Lanzhou, then straight southward to Kunming, and finally

Distribution of solar energy resources in China

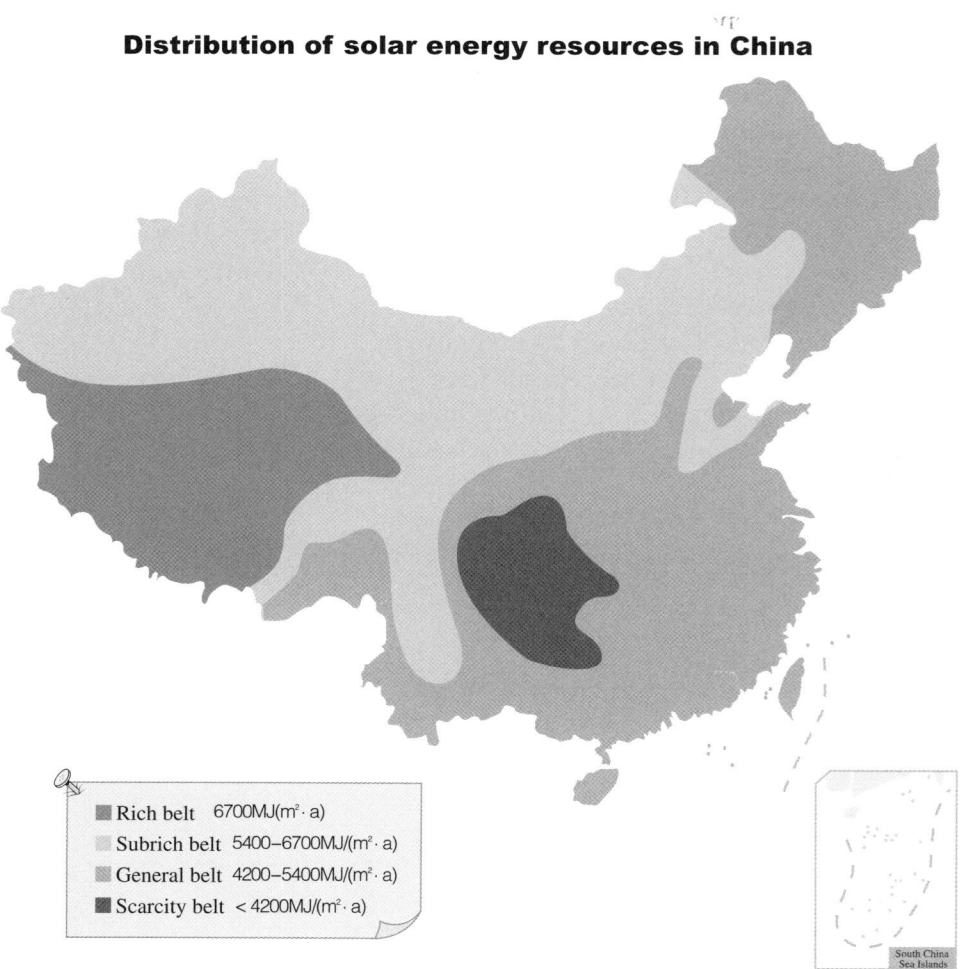

- ■ Rich belt   6700MJ(m²· a)
- Subrich belt   5400–6700MJ/(m²· a)
- General belt   4200–5400MJ/(m²· a)
- ■ Scarcity belt < 4200MJ/(m²· a)

South China
Sea Islands

turning to south Tibet along the Hengduan Mountains. Within the vast regions to the west and north of this isoline, except parts of Xinjiang north of Tianshan Mountain with an annual total of $4.46 \times 10^3 \, \text{MJ}/\text{m}^2$, the majority of the remaining areas boast an annual total of more than $5.86 \times 10^3 \, \text{MJ}/\text{m}^2$.

The western part of Tibet enjoys the most abundant solar energy, with the largest amount up to 2333 kWh/m² (daily radiation being 6.4 kWh/m²), ranking second in the world, second only to the Sahara desert.

Category	Resources	Covered areas
Class I Region (fairly abundant in solar energy resources)	Annual total solar radiation: 6680–8400 MJ/m², equal to daily radiation of 5.1–6.4 kWh/m²	Northern Ningxia, northern Gansu, eastern Xinjiang, western Qinghai and western Tibet, etc.
Class II Region (moderately abundant in solar energy resources)	Annual total solar radiation: 5850–6680 MJ/m², equal to daily radiation of 4.5–5.1 kWh/m²	Northwest Hebei, northern Shanxi, southern Inner Mongolia, southern Ningxia, central Gansu, eastern Qinghai, southeast Tibet and south Xinjiang, etc.
Class III Region (moderately abundant in solar energy resources)	Annual total solar radiation: 5000–5850 MJ/m², equal to daily radiation of 3.8–4.5 kWh/m²	Shandong, Henan, southeast Hebei, southern Shanxi, northern Xinjiang, Jilin, Liaoning, Yunnan, northern Shaanxi, southeast Gansu, southern Guangdong, southern Fujian, northern Suzhou, northern Anhui, and southwest Taiwan, etc.
Class IV Region (relatively poor in solar energy resources)	Annual total solar radiation: 4200–5000 MJ/m², equal to daily radiation of 3.2–3.8 kWh/m²	Hunan, Hubei, Guangxi, Zhejiang, northern Fujian, northern Guangdong, southern Shaanxi, northern Jiangsu, southern Anhui, Heilongjiang, and northeast Taiwan, etc.
Class V Region (scarcest in solar energy resources)	Annual total solar radiation: 3350–4200 MJ/m², equal to daily radiation of 2.5–3.2 kWh/m²	Sichuan Province and Guizhou Province

Chapter 2   Energy Production

Section 1: Comparative analysis of world and China's energy outputs and trends

Oil

In 2009, the global oil output was 3.83 billion tons, a small drop of 2.6% compared with the previous year, the most significant fall since 1982. In contrast, China's total output was 0.19 billion tons, with a year-on-year drop of 0.5%, and a share of 4.9% in the world's total oil output.

In 2010, the global oil output witnessed a strong growth, amounting to 3.91 billion tons and increasing by 2.2% compared with the previous year, the most significant growth margin since 2004. While for China, its total oil output was 0.2 billion tons, with a year-on-year rise of 7.1%, the largest growth margin in history.

Oil outputs all over the world and in China

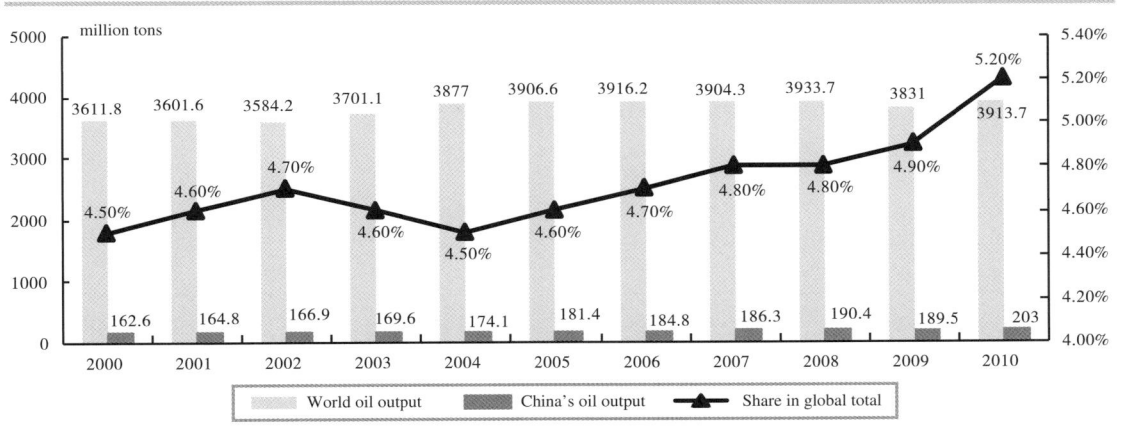

Data source: BP Statistical Review of World Energy, June 2011

Regionally, in 2009, oil outputs in North America, Central and South America, Europe and the former Soviet Union all enjoyed growth; the Middle East, Africa and the Asia-Pacific Region saw a fall; in 2010, except for Europe and the former Soviet Union which witnessed a slight drop in oil production, the remaining regions saw accelerated growth.

Oil outputs in various regions the world over in 2009

	Output (million tons)	Year-on-year growth rate
North America	632.2	2.1%
South America	338.2	0.8%
Europe and former Soviet Union	856.5	0.7%
Middle East	1164.4	−7.4%
Africa	458.9	−5.4%
Asia-Pacific Region	380.8	−1.3%

Oil outputs in various regions the world over in 2010

	Output (million tons)	Year-on-year growth rate
North America	648.2	2.5%
South America	350.0	3.5%
Europe and former Soviet Union	853.3	−0.4%
Middle East	1184.6	1.7%
Africa	478.2	4.2%
Asia-Pacific Region	399.4	4.9%

Data source: BP Statistical Review of World Energy, June 2011

In 2009, the top 10 oil production countries were Russia, Saudi Arabia, the United States, Iran, China, Canada, Mexico, Venezuela, Kuwait and Iraq. Their oil outputs amounted to 2.35 billion tons, a 61.3% share of the world total. Among these countries, many underwent a certain drop in oil production compared with the previous year, and some even witnessed a decline of over 10%, mainly due to the impact of the international financial crisis, which led to a decrease in oil demand, thus the shut-down of oil wells and limited production in many countries one after another.

Oil outputs of the top 10 oil-producing countries in 2009

Country	Output (million tons)	Year-on-year growth rate
Russia	494.2	1.2%
Saudi Arabia	464.7	−9.8%
USA	328.6	7.8%
Iran	201.5	−4.0%
China	189.5	−0.5%
Canada	156.1	−0.4%
Mexico	147.5	−6.5%
Venezuela	124.8	−5.1%
Kuwait	121.7	−11.3%
Iraq	119.8	0.3%

Oil outputs of the top 10 oil-producing countries in 2010

Country	Output (million tons)	Year-on-year growth rate
Russia	505.1	2.2%
Saudi Arabia	467.8	0.7%
USA	339.1	3.2%
Iran	203.2	0.9%
China	203	7.1%
Canada	162.8	4.3%
Mexico	146.3	−0.8%
Venezuela	126.6	1.4%
Kuwait	122.5	0.6%
Iraq	120.4	0.6%

Data source: BP Statistical Review of World Energy, June 2011

In 2010, Russia still remained first among oil-producing countries, followed by Saudi Arabia, the USA, Iran, China, Canada, Mexico, Venezuela, Kuwait and Iraq in descending order. Along with economic recovery, all but Mexico among these top 10 countries witnessed growth in their oil outputs, especially China, hitting a record high in its growth margin.

Coal

In 2009, the global coal output was 6.88 billion tons, increasing by 2.4% compared with the previous year. For China, its output was 2.97 billion tons, with a year-on-year growth rate of 6.1% and a 43.2% share of the world total coal output.

In 2010, the global coal output amounted to 7.27 billion tons, with a salient growth margin of 6.3% compared with the year 2009. China's coal output was 3.24 billion tons, a year-on-year growth rate of 9.0%, accounting for 44.5% of the world total.

Trends of coal output in China and the world over

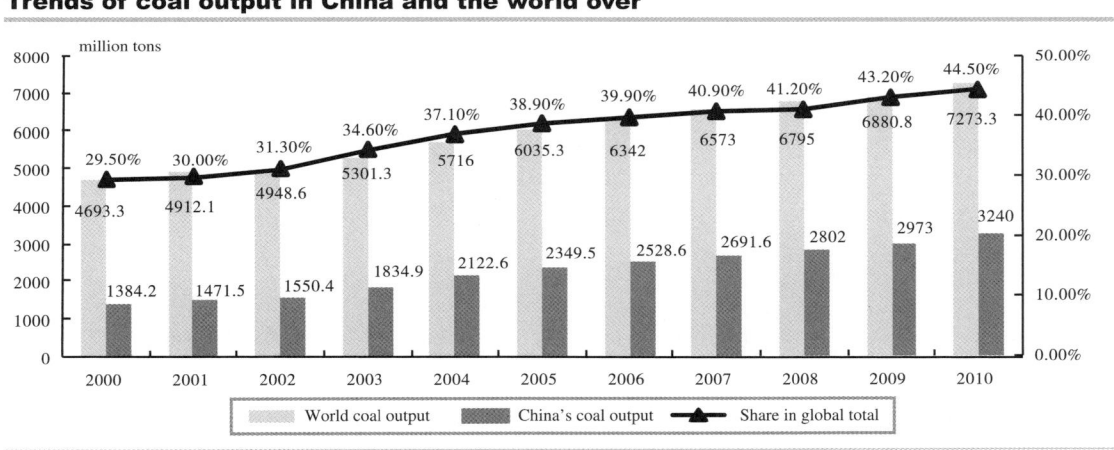

Data source: BP Statistical Review of World Energy, June 2011

Regionally, in 2009, the annual coal output in the Asia-Pacific Region was 4.40 billion tons, with a rise of 8.6% and boasting the rapidest growth. Next was the Middle East with an increase of 0.3%. However, the remaining regions all underwent decline to different degrees: North America dropped by 9.1%, South America by 7.1%, Europe and the former Soviet Union by 6.9%, and Africa by 0.7%.

In 2010, the Asia-Pacific Region, North America, and Europe and Eurasia became the world's largest coal production zones, with their sum total accounting for 94.6% of the global coal output. Among them, the Asia-Pacific Region had a share of 67.2%, North America enjoyed a 15.9% share, and Europe and Eurasia had a share of 11.5%. However, for Africa, Central and South America, and the Middle East as a whole, their sum total only accounted for 5.4% of the world total coal output.

Coal outputs in various regions the world over in 2009

	Output (million tons)	Year-on-year growth rate
North America	1049	−9.1%
South America	82.3	−7.1%
Europe and former Soviet Union	1163.2	−6.9%
Middle East	1.6	0.3%
Africa	253.6	−0.7%
Asia-Pacific Region	4331.1	8.6%

Coal outputs in various regions the world over in 2010

	Output (million tons)	Year-on-year growth rate
North America	1061.8	2.3%
South America	84.5	2.6%
Europe and former Soviet Union	1185.1	2.1%
Middle East	1.6	0.0%
Africa	256.9	1.3%
Asia-Pacific Region	4683.5	8.4%

Data source: BP Statistical Review of World Energy, June 2011

In 2009, the top 10 coal production countries were China, the USA, India, Australia, Russia, Indonesia, South Africa, Germany, Poland and Kazakhstan, whose coal outputs each surpassed 100 million tons, with their sum total amounting to 6.212 billion tons, sharing 89.5% of the global coal output in all.

In 2010, the top two coal production countries were still China and the USA, their total accounting for 63.1% of the world total output. Following were India, Australia, Russia, Indonesia, South Africa, Germany, Poland and Kazakhstan in terms of coal output, among which Indonesia saw a large-scale growth, increasing by 19.4%.

Coal outputs of the top 10 coal production countries in 2009

Country	Output (million tons)	Year-on-year growth rate
China	2973.0	6.1%
USA	975.2	−8.3%
India	556.0	7.8%
Australia	413.2	3.5%
Russia	301.3	−8.3%
Indonesia	256.2	6.6%
South Africa	250.6	−0.8%
Germany	183.7	−4.6%
Poland	135.2	−6.1%
Kazakhstan	100.9	−9.2%

Coal outputs of the top 10 coal production countries in 2010

Country	Output (million tons)	Year-on-year growth rate
China	3240	9.0%
USA	984.6	2.1%
India	569.9	2.5%
Australia	423.9	2.9%
Russia	316.9	4.7%
Indonesia	305.9	19.4%
South Africa	253.8	1.3%
Germany	182.3	−0.5%
Poland	133.2	−1.6%
Kazakhstan	110.8	9.2%

Data source: BP Statistical Review of World Energy, June 2011

Natural gas

In 2009, the global natural gas output declined, for the first time, to 2975.9 billion m³, dropping by 2.8% compared with the previous year. By comparison, China's natural gas output was 85.3 billion m³, with a 2.9% share of the world total and a year-on-year growth rate of 6.2%.

In 2010, this global output of natural gas amounted to 3193.3 billion m³, increasing by 7.3% in comparison to the year 2009, achieving the highest growth margin in history. In contrast, China's natural gas output was 96.8 billion m³, accounting for 3.0% of the world total, with a year-on-year growth rate of 13.5%.

Natural gas outputs in China and the world over

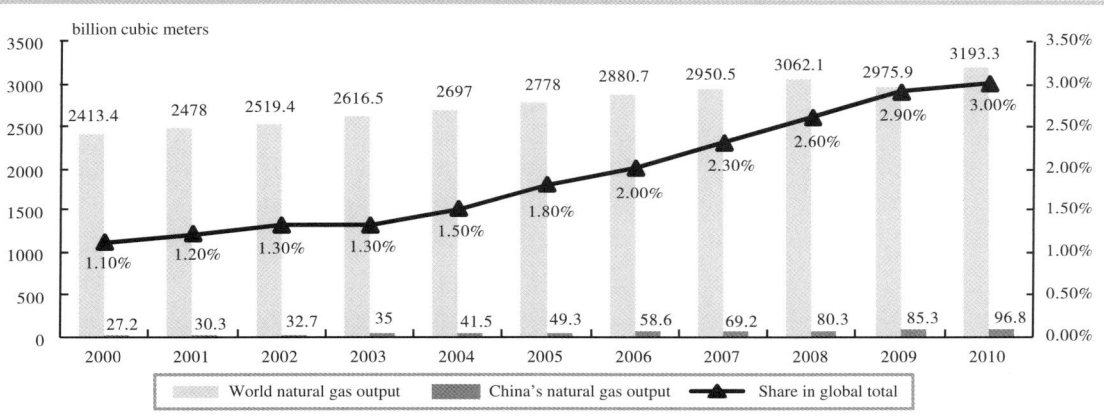

Data source: BP Statistical Review of World Energy, June 2011

Seen regionally, in 2009, except for the Middle East, Asia-Pacific Region and North America which saw an increase, the other regions all witnessed a drop to different degrees, among which Europe and the former Soviet Union underwent the most significant decline, both dropping by 10.4%, followed by Africa with a fall of 4.9%; the next was Central and South America, falling by 3.5%. In contrast, the Middle East and Asia-Pacific Region rose by 6.2% and 4.9% respectively, and North America also enjoyed an increase of 1.4%.

In 2010, along with a strong bounce, the natural gas outputs of all these regions underwent growth, among which the margin for the Middle East and the Asia-Pacific Region ranked top two, with 13.2% and 10.5% respectively.

In 2009, the USA surpassed Russia and became the world's largest natural gas production country. The other 8 countries among the world top 10 were Canada, Iran, Norway, Qatar, China, Algeria, Saudi Arabia and Indonesia in descending order. The total natural gas output of these 10 countries was 1913.9 billion m³, a 64.3% share of the world total.

Natural gas outputs in various regions the world over in 2009		
	Output (billion cubic meters)	Year-on-year growth rate
North America	801.6	0.0%
South America	151.9	−3.6%
Europe and former Soviet Union	969.8	−10.7%
Middle East	407.1	5.9%
Africa	199.2	−5.8%
Asia-Pacific Region	446.4	6.1%

Natural gas outputs in various regions the world over in 2010		
	Output (billion cubic meters)	Year-on-year growth rate
North America	826.1	3.0%
South America	161.2	6.2%
Europe and former Soviet Union	1043.1	7.6%
Middle East	460.7	13.2%
Africa	209.0	4.9%
Asia-Pacific Region	493.2	10.5%

Data source: BP Statistical Review of World Energy, June 2011

In 2010, the USA and Russia still remained in the top two places. The other seven countries were Canada, Iran, Qatar, Norway, China, Saudi Arabia, Indonesia and Algeria. Except for Canada, the other nine countries all witnessed growth in output to different extents.

Natural gas outputs of the top 10 natural gas production countries in 2009		
Country	Output (billion cubic meters)	Year-on-year growth rate
USA	582.8	2.1%
Russia	527.7	−12.3%
Canada	163.9	−7.1%
Iran	131.2	12.8%
Norway	103.7	4.4%
Qatar	89.3	16.0%
China	85.3	6.2%
Algeria	79.6	−7.0%
Saudi Arabia	78.5	−2.4%
Indonesia	71.9	3.2%

Natural gas outputs of the top 10 natural gas production countries in 2010		
Country	Output (billion cubic meters)	Year-on-year growth rate
USA	611	4.7%
Russia	588.9	11.6%
Canada	159.8	−2.5%
Iran	138.5	5.6%
Qatar	116.7	30.7%
Norway	106.4	2.5%
China	96.8	13.5%
Saudi Arabia	83.9	7.0%
Indonesia	82	14.0%
Algeria	80.4	1.1%

Data source: BP Statistical Review of World Energy, June 2011

Section 2: Composition analysis of China's energy generation

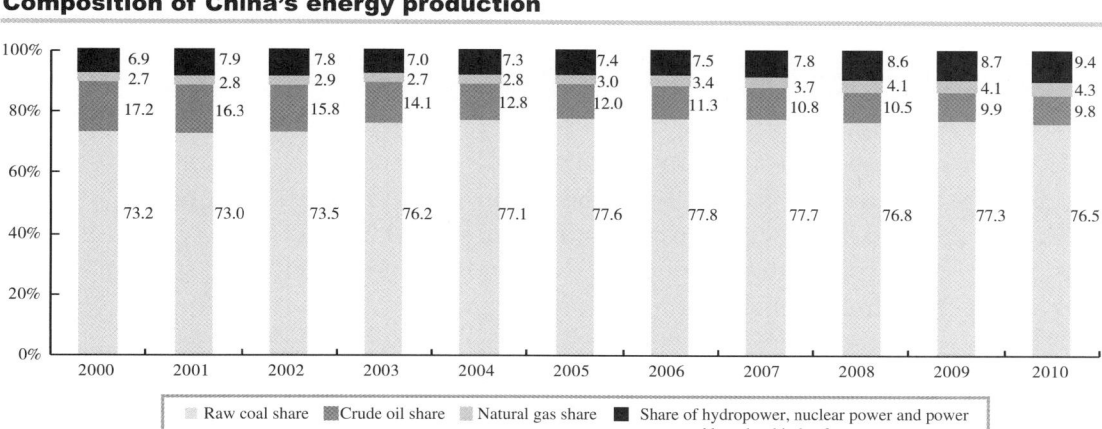

Composition of China's energy production

Data source: China Statistical Yearbook 2011

In 2009, China's primary energy output, when converted to standard coal, amounted to 2.75 billion tons, 5.4% higher than that in the past year; in 2010, the primary energy output, when converted to standard coal, was 2.97 billion tons, rising by 8.1% compared with the previous year. In China, the output of primary energy rose from 21.6 billion tons of standard coal at the end of the Tenth Five-Year period, up to 2.97 billion tons, registering an average annual growth of 6.5%.

In 2009, among the kinds of primary sources of energy in China, raw coal, the major source, witnessed an increase in share to 77.3%. However, the crude oil share fell to 9.9%; the share of natural gas remained at 4.1%; and the share of hydropower, nuclear power and power generated by other sources rose to 8.7%.

In 2010, coal consumption was still the main form of energy consumption in China. The raw coal share declined slightly to 76.5% and the crude oil share also slightly fell to 9.8%, while the share of natural gas increased to 4.3%, and that of hydropower, nuclear power and power generated by other sources rose to 9.4%.

According to the objective specified in the Twelfth Five-Year energy plan, in this five-year period, China shall step up to impel the development of hydropower and nuclear power, and accomplish in a proactive and orderly manner the task of transferring and utilizing renewable energy such as wind power, solar energy, biomass energy and so on. By 2015, the development scales of renewable energy like hydropower, nuclear power and wind energy are predicted to amount to around 250 million kW, 39 million kW and 110 million kW. Meanwhile, by that time, the natural gas share in primary energy is predicted to increase by 4.4%, the hydropower share and nuclear power share in primary energy both by 1.5%, and the share of new energy such as wind power, solar energy, biomass energy and the like by 1.8%.

Section 3: Regional distribution of world and China's energy production

World energy

The distribution of world energy is rather uneven, with the USA ranking first in energy production. In addition, energy production is also significant in some other countries such as China, Russia, Saudi Arabia, India, Canada, Britain, Iran, Indonesia, Mexico, Norway and Australia.

The Middle East is the most important crude oil production zone in the world. In 2009, its output accounted for 30.4% of the world total, yet with a slight fall compared with the previous year, due to the decrease in oil demand caused by the global economic recession. In 2010, this region maintained its leading status, with the output basically remaining at the same level as that in the past year.

Distribution of oil outputs in various regions the world over in 2009

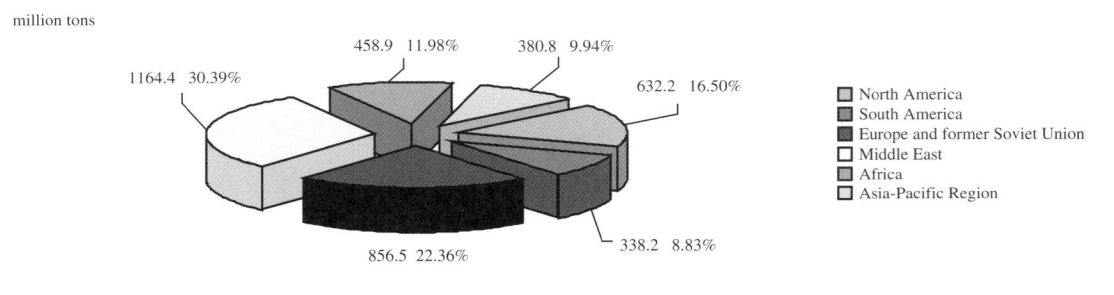

Data source: BP Statistical Review of World Energy, June 2011

Distribution of oil outputs in various regions the world over in 2010

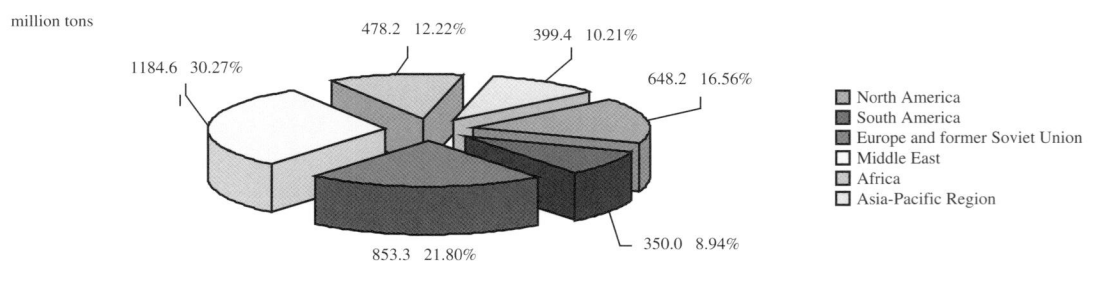

Data source: BP Statistical Review of World Energy, June 2011

The world coal production is mainly concentrated in the Asia-Pacific Region, Europe and the former Soviet Union, and North America.

In 2009, these three major coal production zones boasted a total share of 95.0% of the global output, among which the output of the Asia-Pacific Region ranked first, with a share of 62.9% of the world total.

In 2010, the total output of these three regions accounted for 95.3% of the world total, with the Asia-Pacific Region contributing 64.4%, North America 14.6%, and Europe and the former Soviet Union 16.3%.

Distribution of coal outputs in various regions the world over in 2009

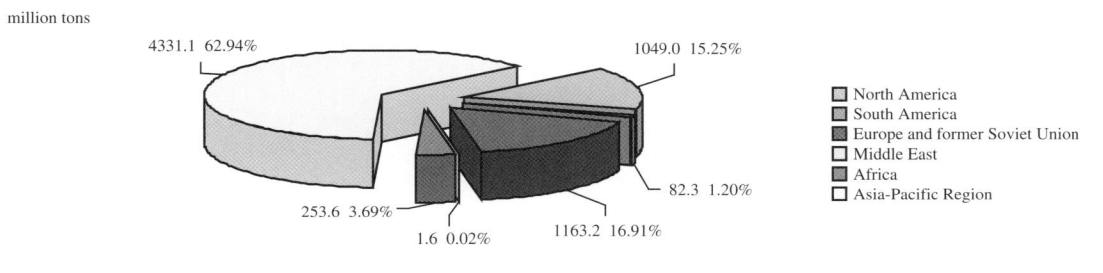

Data source: BP Statistical Review of World Energy, June 2011

Distribution of coal outputs in various regions the world over in 2010

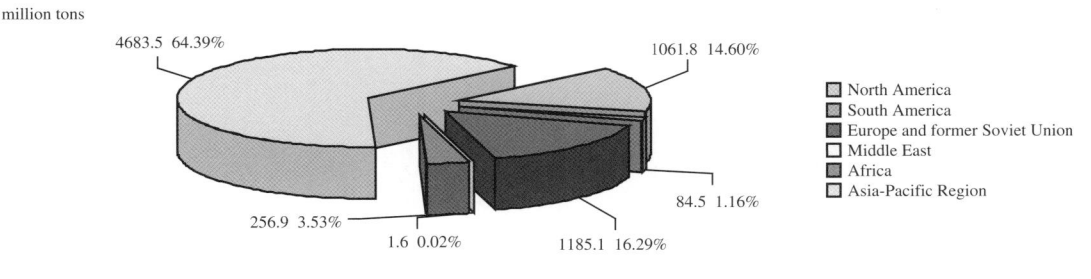

Data source: BP Statistical Review of World Energy, June 2011

The world natural gas production is mainly concentrated in Europe and the former Soviet Union, and North America, whose output in the recent two years both accounted for more than 50% of the global output.

Natural gas output distribution in various regions the world over in 2009

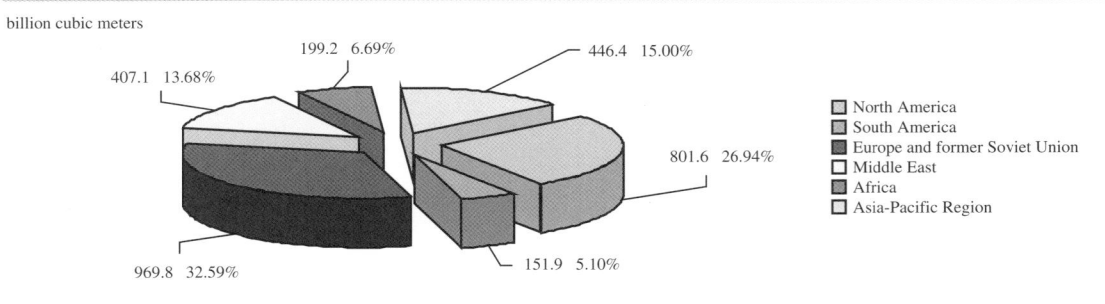

Data source: BP Statistical Review of World Energy, June 2011

Natural gas output distribution in various regions the world over in 2010

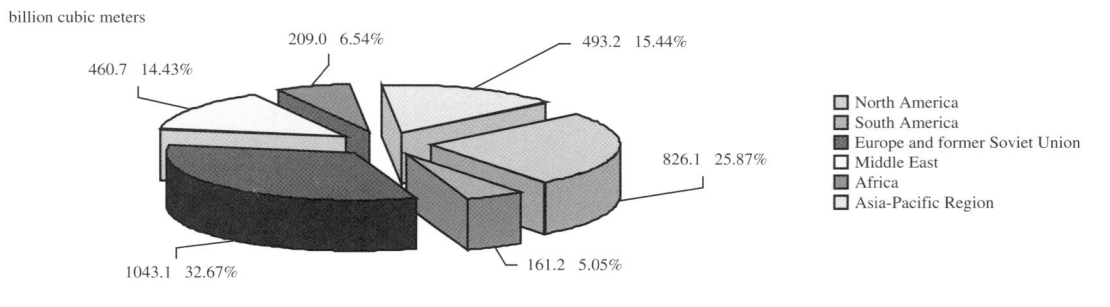

Data source: BP Statistical Review of World Energy, June 2011

China's energy

Coal

China has accelerated the westward movement of its coal production.

In 2009, approximately 2/3 of the raw coal output increase in the country was concentrated within the northwestern regions, Inner Mongolia included: the total output in four provinces and autonomous regions, covering Inner Mongolia, Shaanxi, Ningxia and Xinjiang, witnessed a year-on-year growth of 0.23 billion tons, accounting for 65.9% of national total amount of newly added raw coal output. With the accelerated integration of coal resources, Inner Mongolia obtained an output of raw coal of 0.6 billion tons, with a year-on-year growth of 28.2%, successfully surpassing Shanxi, thus becoming the first big province of coal production in 2009.

In 2010, the coal outputs of Inner Mongolia, Shanxi and Shaanxi provinces still ranked the top three, with the outputs of both Inner Mongolia and Shanxi surpassing 0.7 billion tons for the first time, at 0.78 billion tons and 0.74 billion tons respectively.

China's top 10 provinces in raw coal production in 2009

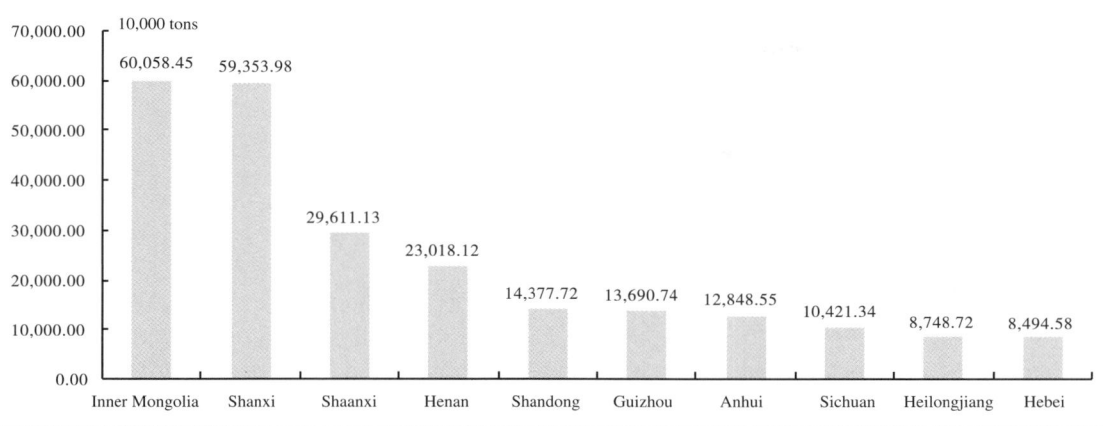

Data source: NBS

China's top 10 provinces in raw coal production in 2010

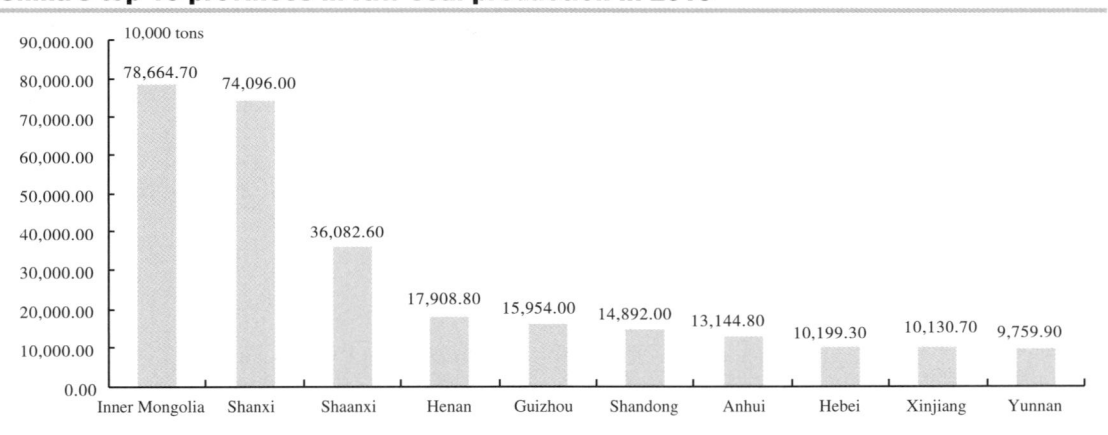

Data source: NBS

In the Eleventh Five-Year period, our country launched key construction projects founding 13 large-scale coal bases in eastern Inner Mongolia, Shendong, northern Shaanxi, western Shandong, Henan, Yunnan-Guizhou, and the two-Huai area, etc. In 2010, these large-sized coal bases achieved a total output of 2.8 billion tons, accounting for about 87.5% of the sum total nationwide.

Oil and natural gas

In 2009, China's top 10 oil-gas fields were, in descending order, Daqing Oilfield, Changqing Oilfield, Shengli Oilfield, Tarim Oilfield, Bohai Sea Oilfield, Xinjiang

Oilfield, Southwest Oil-Gas Field, Yanchang Oil Deposit, Liaohe Oilfield and Tahe Oilfield.

No. 1: Daqing Oilfield (affiliated to PetroChina)

2009: First both in crude oil output and total equivalent of oil and gas.

In 2009, its crude oil output was 40,000,300 tons, with an annual average fall of 3.1%, entering into a production decline period. By contrast, its natural gas output amounted to 3.0 billion m^3 and was expected to witness a further increase and become, hopefully, the second pillar for the development of Daqing Oilfield.

In 2010, it still topped the other oilfields of China. Its production of crude oil remained steady, with an increase of 200,000 tons, achieving a steady crude oil production of 40,000,000 tons for eight consecutive years. On the contrary, the natural gas output underwent a drop of 13.98 million m^3 compared with the previous year.

No. 2: Changqing Oilfield (affiliated to PetroChina)

2009: Third in crude oil output, first in natural gas output and second in total equivalent of oil and gas.

The main operation zone is located in the Shaanxi-Gansu-Ningxia Basin. During the years 2001–2009, the annual average growth rate of crude oil output hit 14.8%; from 2000 to 2009, the average annual growth rate of natural gas output was 29.3%, with the output in 2009 amounting to 18.95 billion m^3, ranking first domestically.

In 2010, the crude oil output amounted to 18.25 million tons, and the natural gas output was 21.1 billion m^3. The total equivalent of oil and gas output during this year was above 35 million tons, a 5-million-ton increase over the past year.

No. 3: Shengli Oilfield (affiliated to Sinopec)

2009: Second in crude oil output and third in the total equivalent of oil and gas.

It lies within the Yellow River delta off the Bohai Sea north of Shandong. From 1999 to 2004, its crude oil output was basically steady. Later on, this output began to grow slowly, and in 2009 amounted to 27,835,000 tons, ranking second in China. As for natural gas production, the peak appeared in 1989, and from then on it began to decline.

In 2010, the crude oil output was 27.34 million tons, dropping by 2% compared with the past year.

No. 4: Tarim Oilfield (affiliated to PetroChina)

2009: Tenth in crude oil output, second in natural gas output and fourth in total equivalent of oil and gas.

The crude oil output has been increasing year by year, up to an all-time new high in 2008. In 2009, however, it slightly fell to 5.54 million tons, ranking tenth domestically. Natural gas, as one of the main forces for the West-East Natural Gas Transmission Project, witnessed a significant increase in its output from 1.36 billion m^3 in 2004 to 18.09 billion m^3, registering an annual average growth rate of 67.9%.

No. 5: Bohai Sea Oilfield (affiliated to CNOOC)

2009: Fourth in crude oil output and fifth in total equivalent of oil and gas.
As the largest offshore oil base of China, this oilfield has seen a year-by-year increase in output during recent years. In 2009, its crude oil output was about 13.5 million tons, ranking fourth in China.

No. 6: Xinjiang Oilfield (affiliated to PetroChina)

2009: Sixth both in crude oil output and total equivalent of oil and gas.
The output of crude oil maintained a steady rapid growth, coming to 10.89 million tons in 2009, thus ranking sixth in China. However, it slightly declined by 10.8% compared with the same period of the previous year, the first drop in recent years. Its natural gas production also basically maintained a rising trend, with 3.6 billion m^3 in 2009, a net increase by 0.2 billion m^3 in comparison with the previous year.

No. 7: Southwest Oil-Gas Field (affiliated to PetroChina)

2009: Seventh in total equivalent of oil and gas.
It is mainly distributed in the Sichuan Basin and Xichang Basin and is now among the list of main-force production zones of natural gas in China. Ever since 2004 when the output of natural gas was 9.777 billion m^3, its natural gas output has witnessed a drastic increase, up to 15.03 billion m^3 in 2009.

No. 8: Yanchang Oil Deposit (affiliated to Yanchang Oil Corporation)

2009: Fifth in crude oil output and eighth in total equivalent of oil and gas.
Located in northern Shaanxi, it has extra-low-permeability reservoirs. In 2009, its crude oil output hit 11.21 million tons, ranking fifth in China.

No. 9: Liaohe Oilfield (affiliated to PetroChina)

2009: Seventh in crude oil output and ninth in total equivalent in oil and gas.
Scattered in the middle and lower reaches of the Liaohe Plains, eastern Inner Mongolia and the beach areas off Liaodong Gulf, since it was put into operation, its crude oil output grew year by year, coming to a peak in 1995 and then falling to about 10 million tons in 2009, thus ranking seventh in China. As for natural gas production, its output appeared to decline year by year since 1995, down to 0.81 billion m^3 in 2009.

No. 10: Tahe Oilfield (affiliated to Sinopec)

2009: Eighth in crude oil output and tenth in total equivalent of oil and gas.
Lying in the northern Tarim Basin, this oilfield sees a rising trend in production year by year. In 2009, its crude oil output and natural gas output amounted to 6.6 million tons and 1.345 billion m^3, respectively.

China's top 10 oil-gas fields in 2009

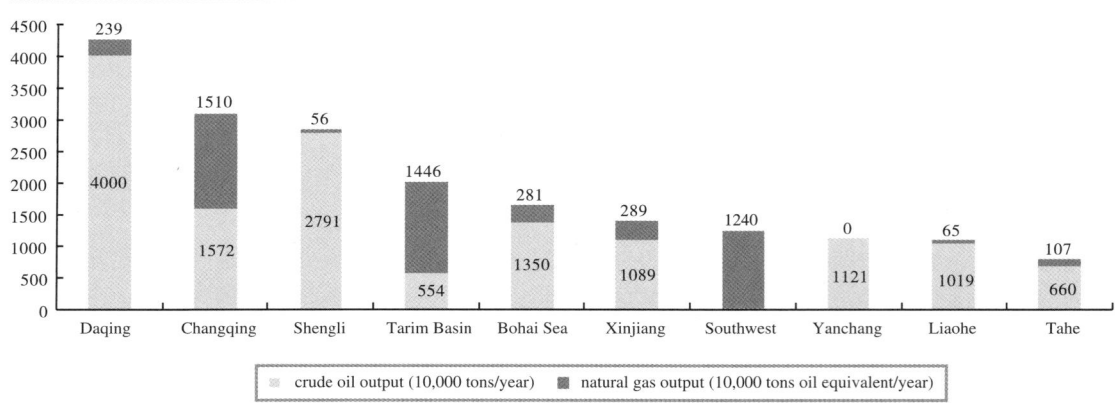

| | crude oil output (10,000 tons/year) | | natural gas output (10,000 tons oil equivalent/year) |

Chapter 3   Energy Consumption

Section 1: Comparative analysis of amount and composition of energy consumption in major countries and China

In 2009, the world consumption of primary energy was 11.36 billion tons of oil equivalent (toe), a slight fall of 1.5% compared with the past year, also the first drop since 1982. The direct cause of this rare decline was the global economic recession.

In 2010, due to the strong pull of the economic recovery, energy of all forms appeared to increase on a large scale: the world energy consumption hit 12.002 billion toe, the largest amount since 1973 and an increase of 5.6% over the previous year.

In 2009, the energy consumption of the industrialized countries, represented by Economic Cooperation (EC), declined by 5.0%, higher than the decrease in their gross domestic product (GDP), the lowest level since 1998. While for the non-EC countries, their energy consumption rose by 2.7%, higher than their GDP growth margin, among which China is the major consumer. The trend that the main consumers of energy consumption are becoming developing countries is continuing.

In 2010, for both EC and non-EC countries, their energy consumption showed a rising trend, with their growth rate above the historical level. The energy consumption of the EC countries rose by 3.5%, the highest growth rate since the year 1984, and that of the non-EC countries increased by 7.5%.

In 2009, the total consumption of primary energy in China witnessed a large-scale growth of 5.2%, accounting for 19.3% of the global total energy consumption. In 2010, China's total consumption of primary energy continued to drastically increase by 11.2%, making up 20.3% of the world total energy consumption, hence surpassing the USA as the largest energy-consuming country.

Trends of the world and China's primary energy consumption

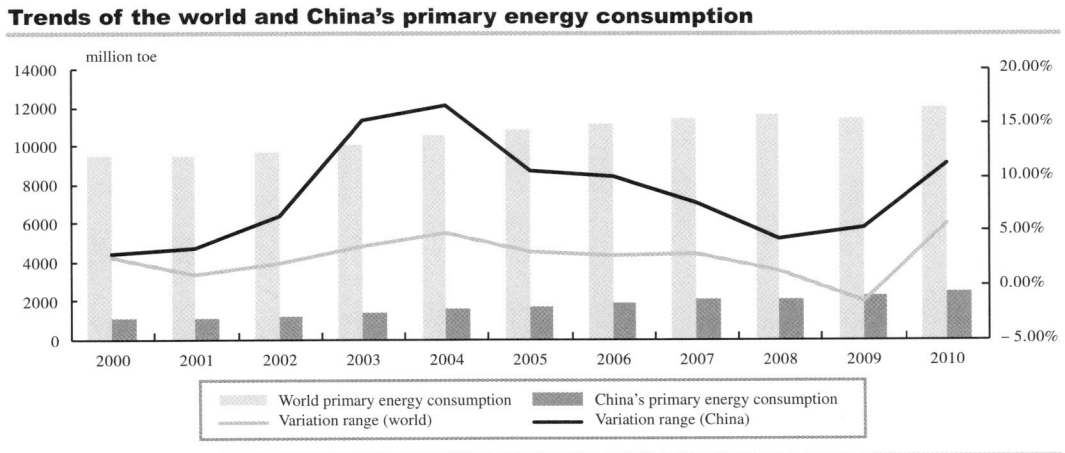

Data source: BP Statistical Review of World Energy, June 2011

In 2009, the composition of world primary energy consumption still had oil, coal and natural gas as the major parts, among which oil accounted for 34.4% of the total primary energy consumption, and coal and natural gas accounted for 29.1% and 23.4% respectively. For other kinds of energy consumption, nuclear power was 5.4%, hydropower 6.5%, and renewable energy 1.2%.

In 2010, for world primary energy consumption, oil was still the kind of energy that took the largest quota, with a 33.6% share of the total consumption. On the other hand, its share has been decreasing for 11 consecutive years. Moreover, coal and natural gas accounted for, respectively, 29.6% and 23.8% of the total primary energy consumption, with nuclear power, hydropower and renewable energy making up 5.2%, 6.5% and 1.2% respectively.

The year 2009 saw a fall in world consumption of oil, natural gas and nuclear energy: oil consumption declined by 1.7%; natural gas dropped by 2.1%, hence making it the kind of fuel with the rapidest decrease; nuclear power reached its consecutive third year of decline, with a fall of 1.3%.

In 2010, various forms of energy consumption the world over witnessed a strong growth, with the growth margin surpassing the average level of the past 10 years.

In 2009, world coal consumption remained steady, thus making it the year of the least variation since 1999. Besides that, only hydropower and other kinds of substitutable energy realized an increase, among which hydropower rose by 1.5%, making it the kind of energy with the highest growth rate in 2009; other forms of renewable energy continued to maintain their momentum of rapid growth.

In 2010, world natural gas consumption increased by 7.4% compared with the year 2009, hitting an all-time high in growth rate; the growth rate of coal consumption hit 7.6%, above the average level of the previous 10 years; oil consumption underwent a strong recovery with its growth rate up to 3.1%, the largest increase since 2004; for hydropower consumption, that figure was 5.3%; and renewable energy witnessed a double-digit growth rate of 15.4%.

World energy consumption composition

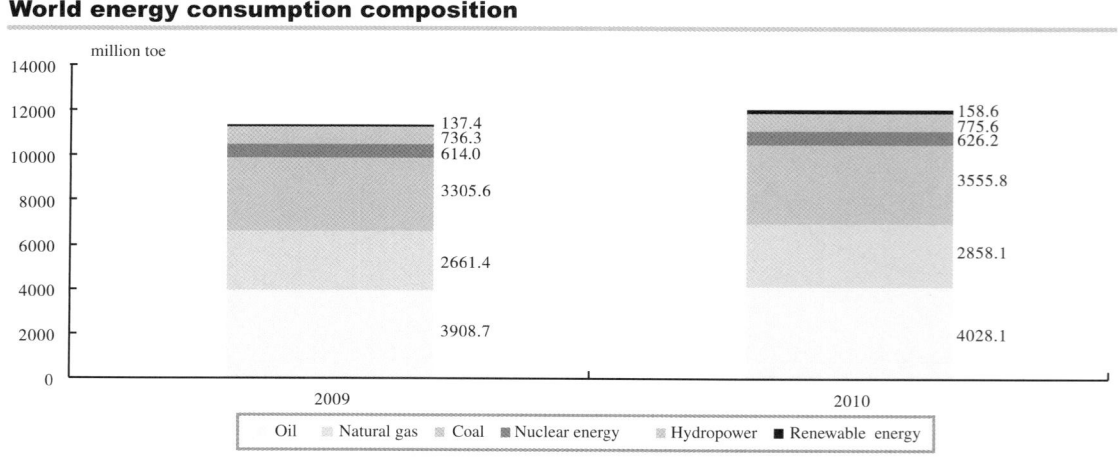

Data source: BP Statistical Review of World Energy, June 2011

In terms of region, in the Asia-Pacific Region, coal occupies the dominant position; in Europe and Eurasia, the major fuel is natural gas; in the remaining regions, the quota of oil consumption ranks first.

Energy consumption composition of various regions the world over in 2009

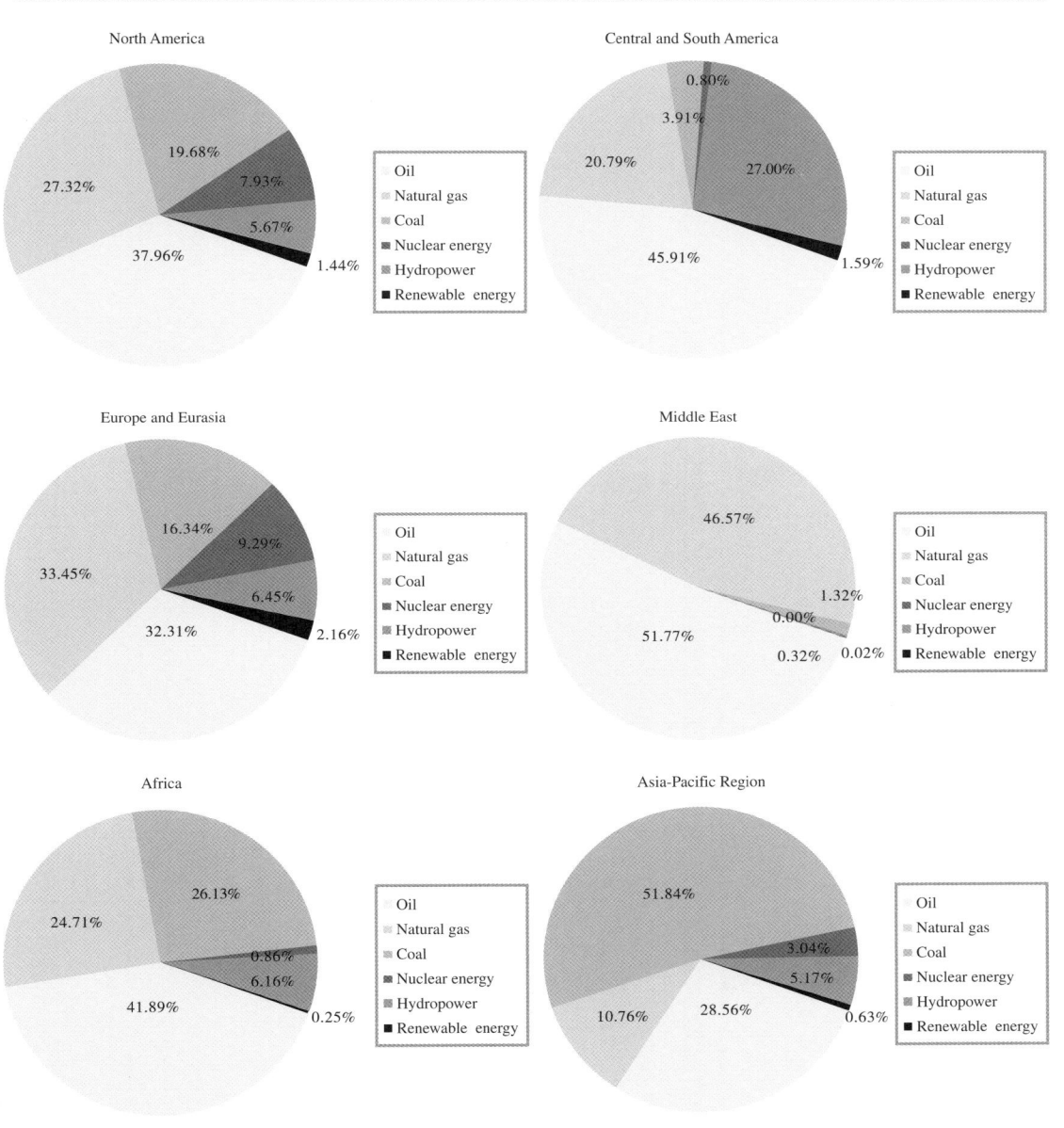

Data source: BP Statistical Review of World Energy, June 2011

Energy consumption composition of various regions the world over in 2010

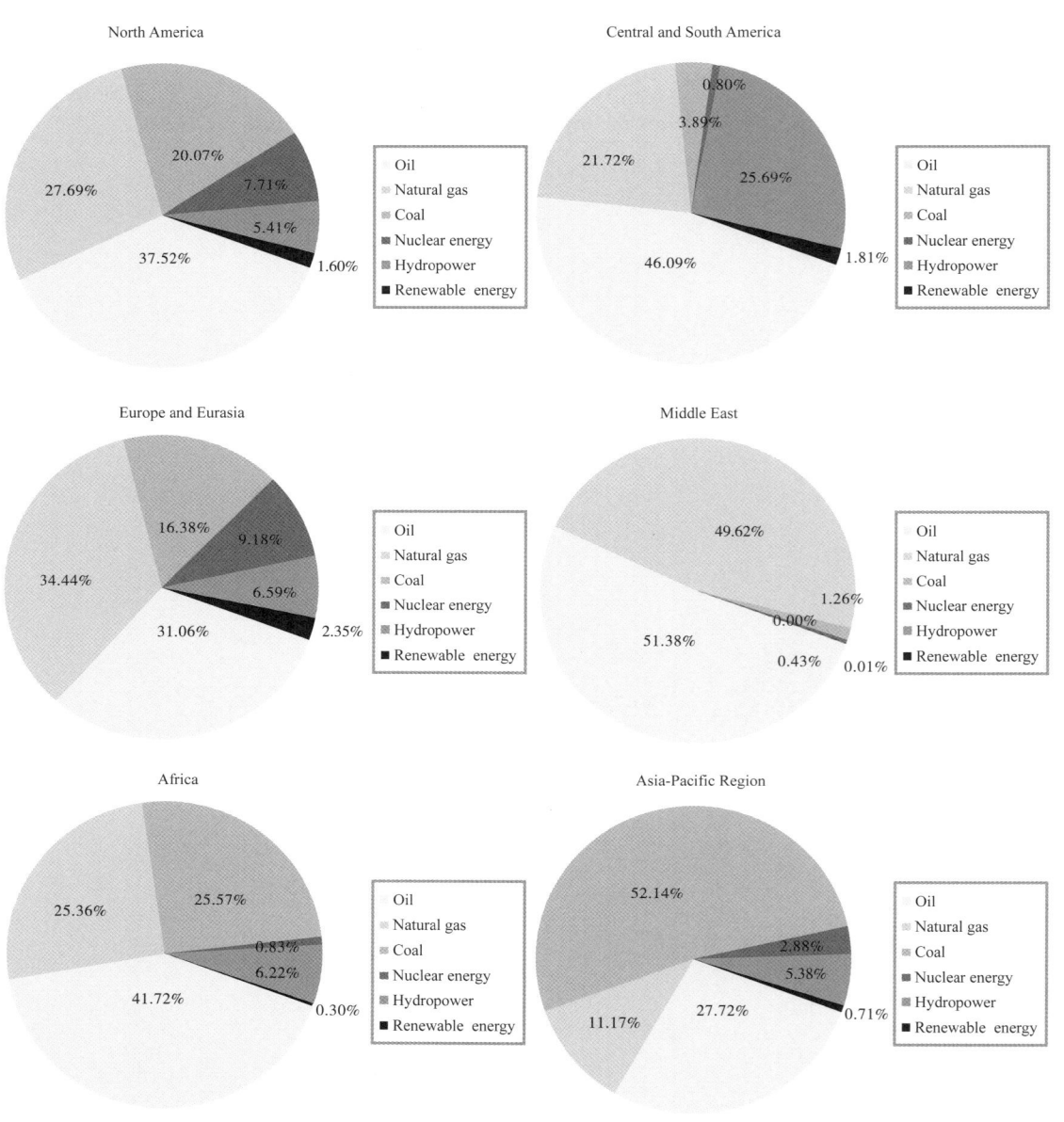

North America

20.07%
7.71%
27.69%
5.41%
37.52%
1.60%

Oil
Natural gas
Coal
Nuclear energy
Hydropower
Renewable energy

Central and South America

0.80%
3.89%
21.72%
25.69%
46.09%
1.81%

Oil
Natural gas
Coal
Nuclear energy
Hydropower
Renewable energy

Europe and Eurasia

16.38%
9.18%
34.44%
6.59%
31.06%
2.35%

Oil
Natural gas
Coal
Nuclear energy
Hydropower
Renewable energy

Middle East

49.62%
1.26%
0.00%
51.38%
0.43%
0.01%

Oil
Natural gas
Coal
Nuclear energy
Hydropower
Renewable energy

Africa

25.36%
25.57%
0.83%
6.22%
41.72%
0.30%

Oil
Natural gas
Coal
Nuclear energy
Hydropower
Renewable energy

Asia-Pacific Region

52.14%
2.88%
5.38%
11.17%
27.72%
0.71%

Oil
Natural gas
Coal
Nuclear energy
Hydropower
Renewable energy

Data source: BP Statistical Review of World Energy, June 2011

In terms of country, although the energy consumption differs between different countries, their energy structure seems to follow certain rules. For developing countries, most of them consume coal as the major source of energy, while the energy structure of developed countries shows that they normally consume oil as the principal source of energy, its consumption accounting for about 40% of the total energy consumption.

Composition of energy consumption in major countries in 2009 (unit: million toe)

Country	Oil	Natural gas	Coal	Nuclear energy	Hydropower	Renewable energy
USA	37.80%	26.70%	22.50%	8.60%	2.80%	1.50%
Canada	31.10%	27.20%	7.50%	6.50%	26.70%	1.10%
France	35.90%	15.60%	4.00%	38.00%	5.40%	1.10%
Germany	37.00%	22.80%	23.30%	9.90%	1.40%	5.50%
Italy	44.60%	38.30%	7.80%	—	6.60%	2.70%
Britain	36.60%	38.30%	14.50%	7.70%	0.60%	2.30%
Russia	20.60%	53.60%	14.00%	5.70%	6.10%	0.00%
Japan	42.00%	16.60%	23.00%	13.70%	3.50%	1.10%
South Korea	43.50%	12.90%	29.00%	14.10%	0.30%	0.20%
India	31.50%	9.60%	52.20%	0.80%	5.00%	1.00%
China	17.70%	3.70%	71.20%	0.70%	6.40%	0.30%
World total	34.40%	23.40%	29.10%	5.40%	6.50%	1.20%

Data source: BP Statistical Review of World Energy, June 2011

Composition of energy consumption in major countries in 2010 (unit: million toe)

Country	Oil	Natural gas	Coal	Nuclear energy	Hydropower	Renewable energy
USA	37.20%	27.20%	23.00%	8.40%	2.60%	1.70%
Canada	32.30%	26.70%	7.40%	6.40%	26.20%	1.10%
France	33.10%	16.70%	4.80%	38.40%	5.70%	1.30%
Germany	36.00%	22.90%	24.00%	10.00%	1.40%	5.80%
Italy	42.50%	39.80%	8.00%	0.00%	6.50%	3.20%
Britain	35.20%	40.40%	14.90%	6.70%	0.40%	2.40%
Russia	21.40%	53.90%	13.60%	5.60%	5.50%	0.00%
Japan	40.20%	17.00%	24.70%	13.20%	3.80%	1.00%
South Korea	41.40%	15.10%	29.80%	13.10%	0.30%	0.20%
India	29.70%	10.60%	52.90%	1.00%	4.80%	1.00%
China	17.60%	4.00%	70.50%	0.70%	6.70%	0.50%
World total	33.60%	23.80%	29.60%	5.20%	6.50%	1.30%

Data source: BP Statistical Review of World Energy, June 2011

Energy consumption composition in China

Abundant in coal resources, China is the world's largest coal consumer, its coal consumption making up 27% of the world total coal consumption. The reality of being rich in coal, scarce in natural gas has led to a situation where over a long period coal has been playing a highly significant role in China's primary energy consumption, making China among a handful of large energy-consuming countries with coal as the mainstay.

In 2009, coal accounted for 71.2% of China's total energy consumption. Compared with the world energy structure, the share of coal consumption in China was relatively large, 42.1 percentage points higher than the world average level, while for oil and natural gas, their shares were lower than the world average level by 16.7 percentage points and 19.7 percentage points respectively.

In 2010, coal still was the principal source of energy, accounting for 70.5% of the world total of energy consumption.

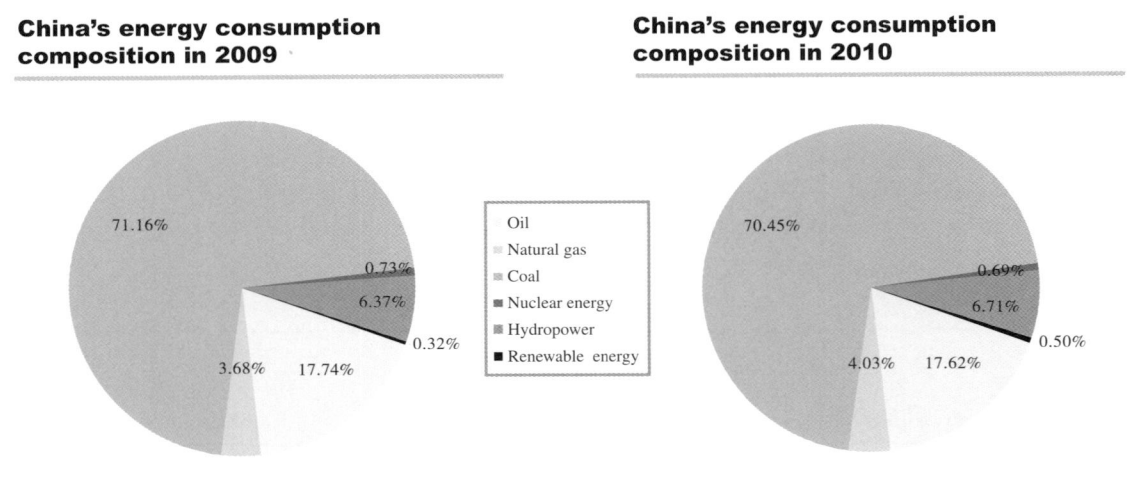

China's energy consumption composition in 2009

China's energy consumption composition in 2010

Energy consumption composition in the USA

Different from the developing countries, the energy structure of the developed countries usually takes oil as the first choice. Their oil consumption accounts for about 40% of the world total energy consumption, with the case of the USA all the more typical.

In 2009, America's oil consumption made up 37.8% of the total energy consumption; in 2010, its oil consumption still occupied the largest share, amounting to 37.2%.

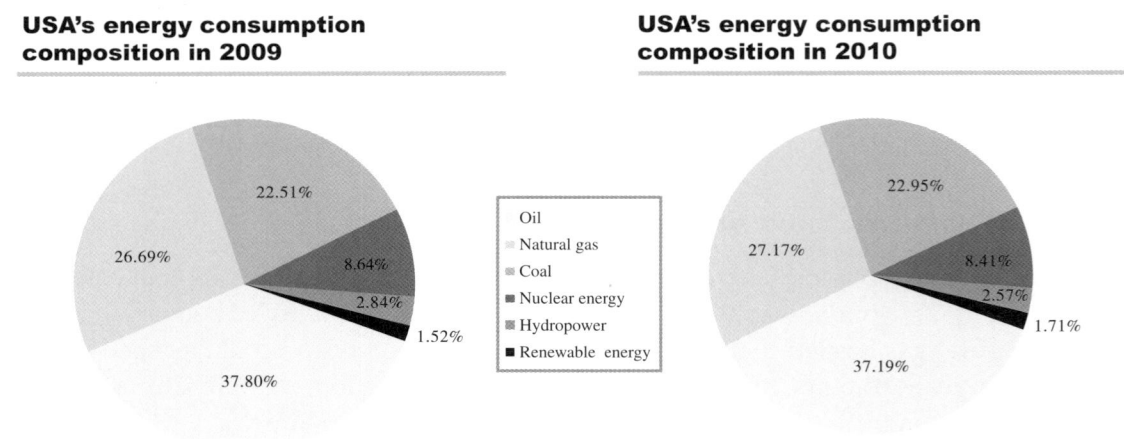

USA's energy consumption composition in 2009

USA's energy consumption composition in 2010

Energy consumption composition in Russia

For Russia, the reliance upon oil is relatively low, but the share of natural gas is comparatively high.

In 2009, Russia's natural gas consumption had a 53.6% share of the total energy consumption; in 2010, this share changed into 53.9%.

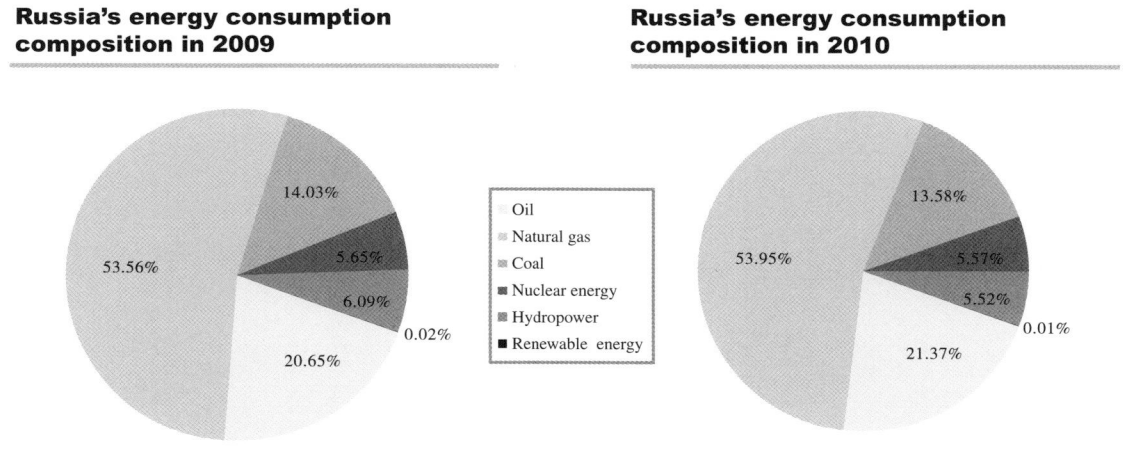

Russia's energy consumption composition in 2009

Russia's energy consumption composition in 2010

Oil
Natural gas
Coal
Nuclear energy
Hydropower
Renewable energy

Energy consumption composition in India

India is scarce in oil, and like China, it relies on coal as the major source of energy. Furthermore, oil resources have gained a more and more important place in energy consumption.

In 2009, India's coal consumption accounted for 52.9% of the total energy consumption, its share of oil consumption being 29.7%; in 2010, the share of coal consumption was 52.2%, and that of oil consumption rose slightly to 31.5%.

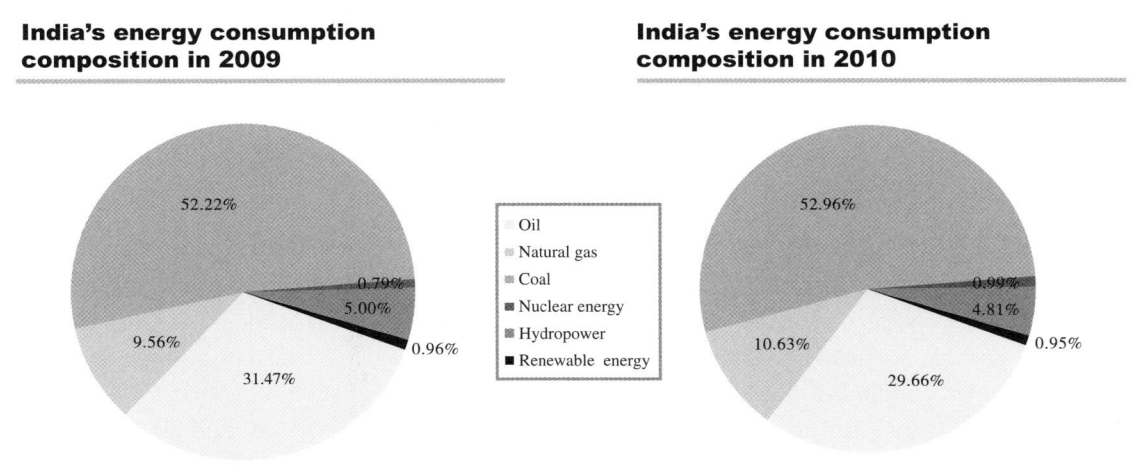

India's energy consumption composition in 2009

India's energy consumption composition in 2010

Oil
Natural gas
Coal
Nuclear energy
Hydropower
Renewable energy

Energy consumption composition in Japan

Japan is a country relatively scarce in energy, and thus has a high degree of dependence on energy imports. In recent years, China, Japan has gradually lessened its oil consumption and adjusted its energy structure through increasing the shares of hydropower and nuclear power, with oil still remaining the main part of Japan's energy consumption.

In 2009, Japan's oil consumption accounted for 42.0% of the total energy consumption; in 2010, the share of oil consumption slightly dropped to 40.2%.

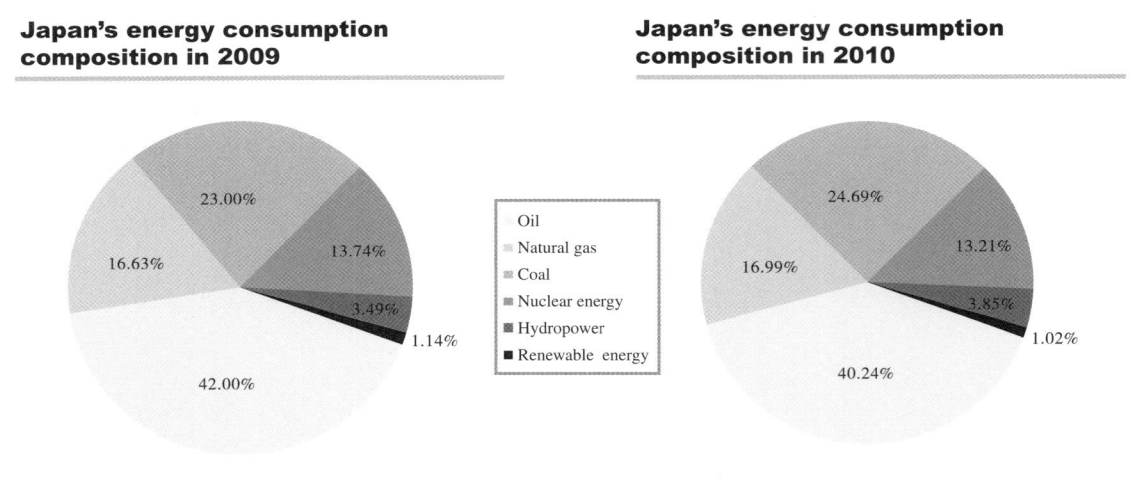

Japan's energy consumption composition in 2009

Japan's energy consumption composition in 2010

- Oil
- Natural gas
- Coal
- Nuclear energy
- Hydropower
- Renewable energy

Section 2: Trend analysis of world and China's energy consumption

IEA World energy outlook

On November 9, 2010, the report *World Energy Outlook* issued by the International Energy Agency (IEA) claimed that the emerging economy led by China and India would push the global energy demand in the coming 25 years. China's demand for energy would rise by 75% during the years from 2008 to 2035. By 2035, the share of China's energy demand in global energy demand would have risen from the current 17% to 22%.

BP energy outlook 2030

On January 22, 2011, the report on energy trends *Energy Outlook 2030*, issued by BP corporation, pointed out that in the next 20 years, the primary energy consumption would increase by 40% approximately, with 93% of this increase contributed by non-EC countries. The quota of non-EC countries' energy demand would drastically grow from slightly higher than 1/2 to 2/3.

BP's "baseline situation" prediction claims that, in the period between 2010 and 2030, the annual average growth rate of world primary energy demand will be 1.7%, which will slightly slow down by 2020. The energy consumption of non-EC countries will increase by 68% by 2030, at an annual average growth rate of 2.6%, the growth accounting for 93% of the global energy growth. In comparison, by 2030, the annual average growth rate for EC countries will be 0.3%; moreover, from 2020 onwards, the energy consumption per capita in EC countries will show a declining tendency, a yearly drop of 0.2%.

It predicts that the market shares of coal and oil will fall and that natural gas will be the kind of fuel with the rapidest growth. Meanwhile, diversification of energy supply will be further strengthened, and such forms of non-fossil energy as nuclear energy, hydropower and renewable energy are expected to become the major source of supply growth for the first time.

In the 20 years to come, the global demand for liquid fuel will all come from non-EC emerging economies, 2/3 of which is contributed by the energy demand growth in non-EC Asian countries. In contrast, the overall demand of the EC countries for oil and other liquid fuels reached its peak in 2005, and will approximately drop to the level of the year 1990 by 2030.

Natural gas is a fossil fuel with a rapid growth, which is over three times oil's expected annual growth rate of 2.1%. The main driving forces for the growth in natural gas demand are non-EC countries and the global power generation field. Among them, the countries and regions with the rapidest growth will be China, India, Brazil and the Middle East.

Coal will see an annual growth rate of 1.2%, and by 2030, except for biofuel, it will be utilized more to the same degree as oil. The risks that EC countries face will be strongly driven by carbon policies.

China's energy consumption

China's coal consumption will continue to maintain a moderate growth. However, under the guiding ideology of the overall strategy of "less use of oil and more development of natural gas", especially influenced by the tasks of energy saving and emission reduction in the Twelfth Five-Year period, the share of coal in China's energy consumption composition is expected to drop from 70.7% to around 50.7% during the years 2010–2020. In addition, the shares of natural gas, hydropower, nuclear power and wind power will also witness an increase.

Natural gas consumption will continue to maintain a rapid growth, and its consumption area will further expand in that the place of production and its surroundings, the Bohai Sea Rim, Changjiang River Delta and the economically developed areas off the southeast coast, will become the primary consumption area, with its consumption structure continuing a diversified development.

The power demand of major energy-consuming industries will see a rebound and maintain a steady growth.

In 2011, the Twelfth Five-Year Plan for renewable energy will be in full operation; the development plan for seven strategic emerging industries will also be comprehensively carried out. Besides, new energy and renewable energy will welcome a brand-new period of opportunities for development.

Section 3: Analysis of China's energy consumption and utilization

In 2009, energy supply and demand in China experienced highs and lows, from the ensuing bottoming out after the international financial crisis to a gradual recovery along with the economic pick-up, as well as the change in national economy to obtain a steady growth. During the years 2004–2009, the growth rate of China's total energy consumption fell from 16.2% to 5.2%.

In 2010, under the strong pull of the economic recovery, energy demand witnessed a large-scale growth, with a year-on-year growth of 11.2% in total energy consumption.

China's primary energy consumption

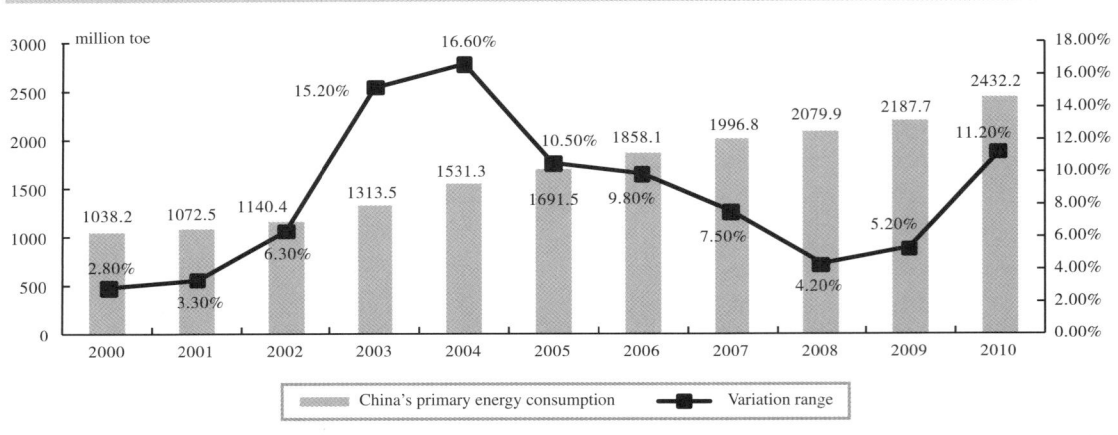

Data source: BP Statistical Review of World Energy, June 2011

In 2009, China's total primary energy consumption accounted for 11.5% of the world total, ranking second only to that of the USA; in 2010, China surpassed the USA as the biggest energy-consuming country in the world.

In China's energy consumption, coal has always had a share of over 2/3, and this pattern with coal as the major consumption item will last for a long period. During the period 2009–2010, in China's primary energy composition, the share of coal rose from 67.8% to 70.4%; that of oil dropped from 23.2% to 17.9%; the natural gas share increased from 2.4% to 3.9%; and the share of hydropower, nuclear power and wind power in all rose from 6.7% to 7.8%, which had undergone a fall in 2003 and rebounded to rise from then on.

In 2010, China's total primary energy composition increased from 2.96 billion tons by the end of the Tenth Five Year period to 3.2 billion tons of standard coal, with an annual average growth rate of 6.3%. Among this total, the share of coal dropped to 68.0%; that of oil rose to 19.0%; the natural gas share increased to 4.4%; and the share of hydropower, nuclear power and wind power in all increased to 8.6%.

Shares of various energy consumption items in China

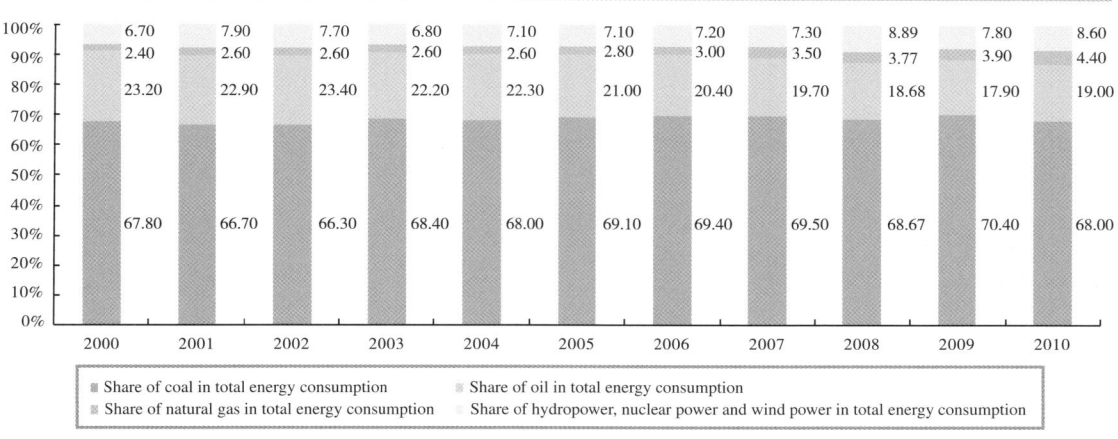

■ Share of coal in total energy consumption ■ Share of oil in total energy consumption
■ Share of natural gas in total energy consumption ■ Share of hydropower, nuclear power and wind power in total energy consumption

Data source: China Statistical Yearbook

With the growth in energy consumption in China, its per capita energy consumption has witnessed a gradual increase, yet remains far lower than that of the developed countries. In 2009, China's per capita energy consumption was 2303 kg of standard coal, which, when converted to standard oil, was 1.6 tons of standard oil, about 1/5 of that for the USA. In 2010, China's per capita energy consumption was 2380 kg of standard coal, increasing by 32% compared with that in 2005.

Per capita energy consumption

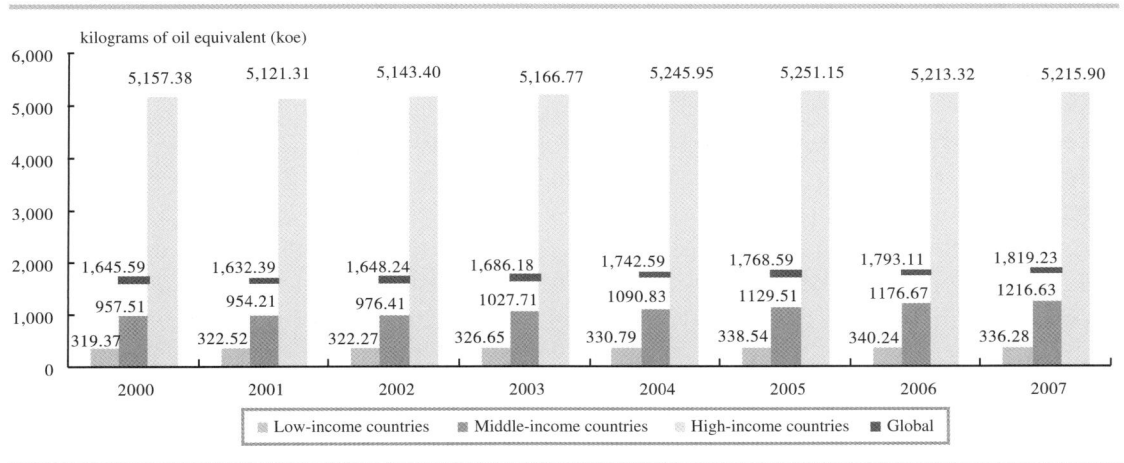

■ Low-income countries ■ Middle-income countries ■ High-income countries ■ Global

Data source: World Bank

China's per capita energy consumption

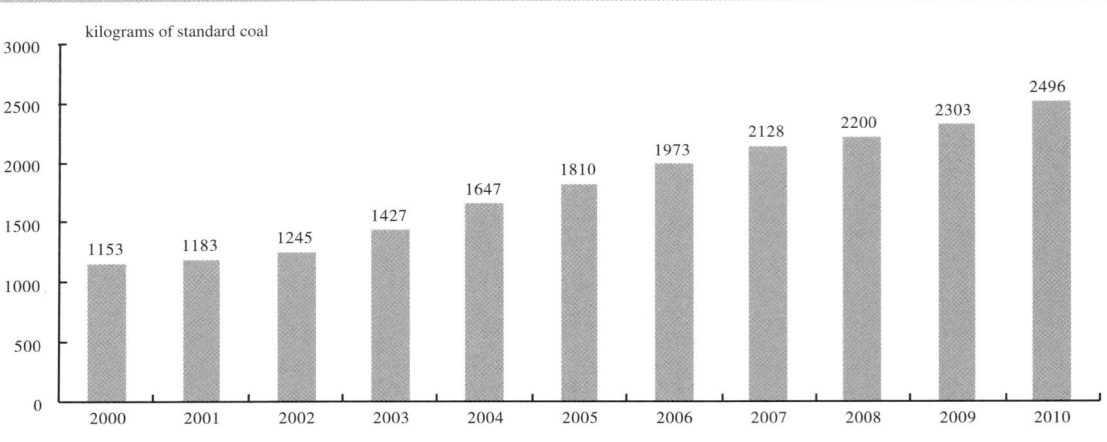

Data source: Energy Statistical Yearbook

The concentration ratio of China's energy consumption has been further reinforced: a rapid growth in energy consumption has been witnessed in six leading high-energy-consuming industries, i.e. petroleum processing, coking and nuclear fuel processing, chemical raw material and chemical manufacturing, non-metal mineral product manufacturing, smelting and rolling processing of ferrous metals, smelting and rolling processing of non-ferrous metals, and production and supply of electric and thermal power. In 2009, the total energy consumption of these six industries was 3.066 billion tons of standard coal, 2.1 times that in 2000, and this growth margin was higher than that of energy consumption. The energy consumption share of these six leading high-energy-consuming industries in China's total energy consumption also keeps rising, from 47.5% in 2000 to 51.7% in 2009, exceeding half of China's total energy consumption.

Among the six leading high-energy-consuming industries, the industry of smelting and rolling processing of ferrous metals consumes the largest amount of energy, and also has the highest growth rate in energy consumption. In 2009, its energy consumption was 0.564 billion tons of standard coal, making up 35.58% of the total energy consumption of the six high-energy-consuming industries, 2.9 times that in 2000. By comparison, the industry of smelting and rolling processing of non-ferrous metals is claimed to have the least energy consumption, yet with a relatively rapid growth. In 2009, its energy consumption was 0.11 billion tons of standard coal, 2.8 times that in 2000.

Energy consumption of six leading high-energy-consuming industries

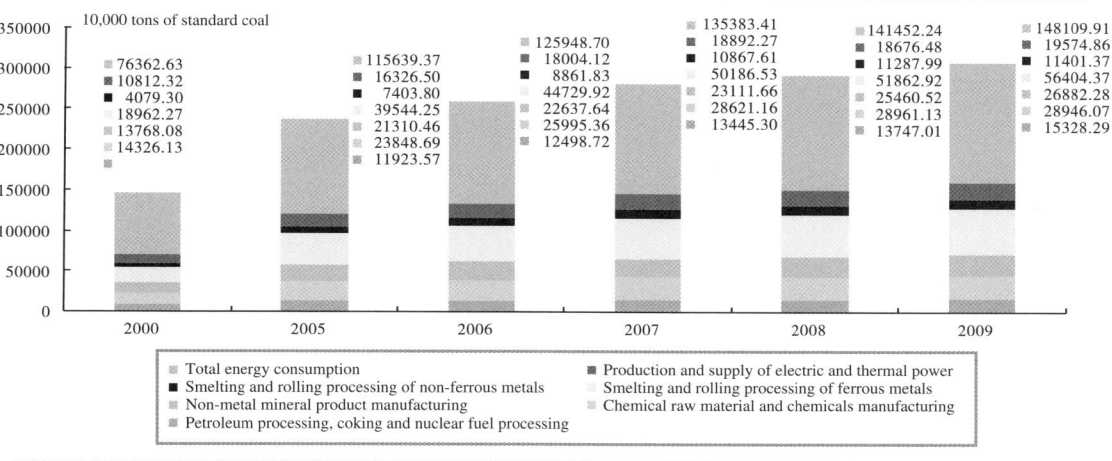

Data source: China Statistical Yearbook

China's energy efficiency represents a rising trend, yet still remains relatively low. In recent years, our country has made great efforts to implement various policies and measures to save energy and reduce emissions, proactively adjust the industrial structure, accelerate the phasing out of backward production capability, leading to a continuous decline in energy consumption intensity, thus supporting the steady and relatively rapid development of national economy with a comparatively low growth rate of energy consumption.

In the Eleventh Five-Year period, the energy consumption for every 10,000 yuan of GDP dropped from 1.276 tons of standard coal in 2005 to 1.034 tons of standard coal in 2010, falling by 2.7%, 5.0%, 5.2%, 3.6% and 4.0% compared to the previous year, with a cumulative drop of 20.5% and an annual average fall of 4.1%.

Trend of GDP energy consumption in China

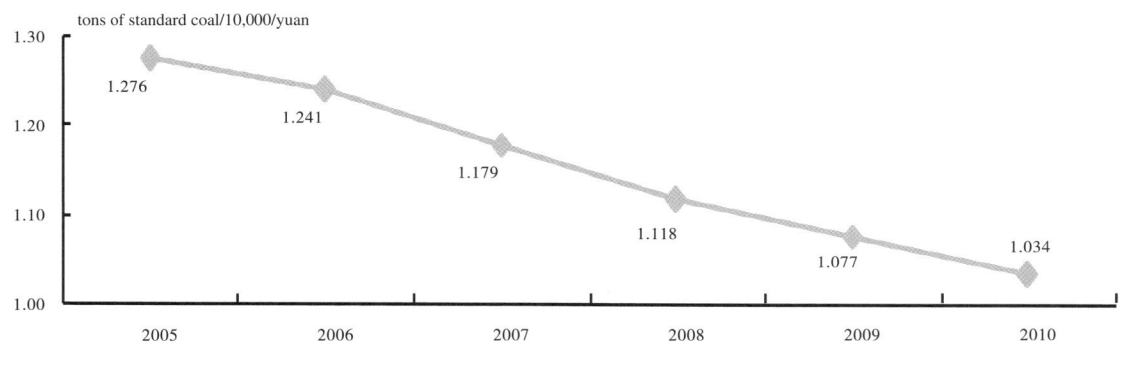

Data source: China Statistical Yearbook

Chapter 4   Energy Trade

Section 1: Price and trade distribution of major types of energy in the world

Oil

In 2009, although the global oil demand always had a negative growth, yet as the OPEC cutback effect was gradually emerging, along with dollar depreciation and the involvement of large-holding international speculative capital, international oil prices generally speaking, witnessed a tendency to rebound upwards after undergoing a small-scale decline and continue to go up constantly.

In early 2009, oil prices continued the declining tendency from the end of the previous year, and kept going down. During the period from mid-August to early October, oil prices fluctuated between 65–75 dollars/barrel. On October 14, international oil prices surpassed the mark of 75 dollars/barrel. From then onwards and till the end of November, and prices remained between the range of 75–80 dollars/barrel, and ended up at 79.4 dollars/barrel by the end of the year. WTI's average price for the year was 61.9 dollars/barrel, dropping by 38% compared with the same period of the previous year.

In 2010, with the gradual recovery of the world economy, the international oil prices for the year as a whole witnessed a fluctuating increase. Influenced by the debt crisis in Europe, the change in the US dollar exchange rate and other factors, international oil prices once again underwent a drastic fluctuation. However, thanks to the generally sound momentum of the global economy, oil prices also maintained an upward trend during the year as a whole.

In early 2010, oil prices continued the rising tendency from the end of the previous year, with a high-level start. In May, the European debt crisis broke out, the global commodity market and share market experienced a large-scale adjustment and oil prices declined to around 70 dollars/barrel. In early August, the international oil prices represented a continuing tendency of concussive upstream from the end of July, after then fell to deep bottom accompanied with a series of bad news. Starting from late November, a series of good news on economic growth propped up oil prices for them to rebound. On December 22, due to a fall in crude oil inventory in the USA and an increase in economic growth rate during the third quarter, international oil prices cracked once again through the 90 dollars/barrel mark after a period of two years, ending up at 91.4 dollars/barrel. WTI's average price for the year was 79.5 dollars/barrel, increasing by 29% over the previous year.

International oil price trends during the years 2001–2010

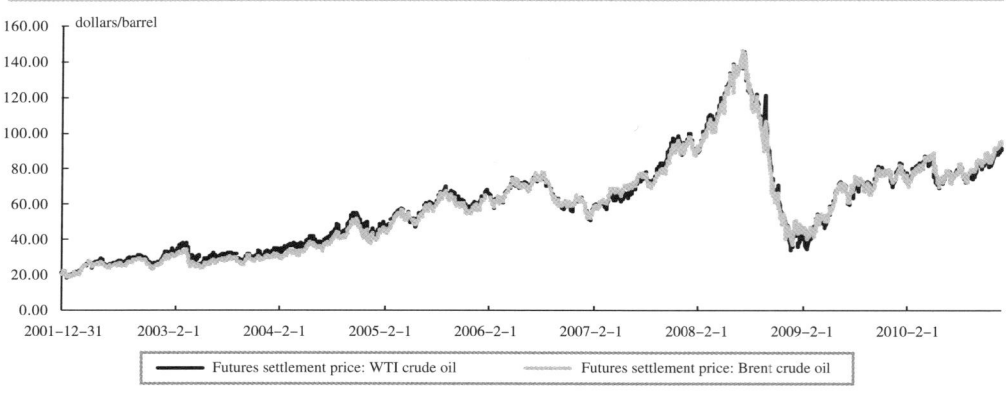

Futures settlement price: WTI crude oil   Futures settlement price: Brent crude oil

Data source: Nymex, IPE

International oil price trends in 2009

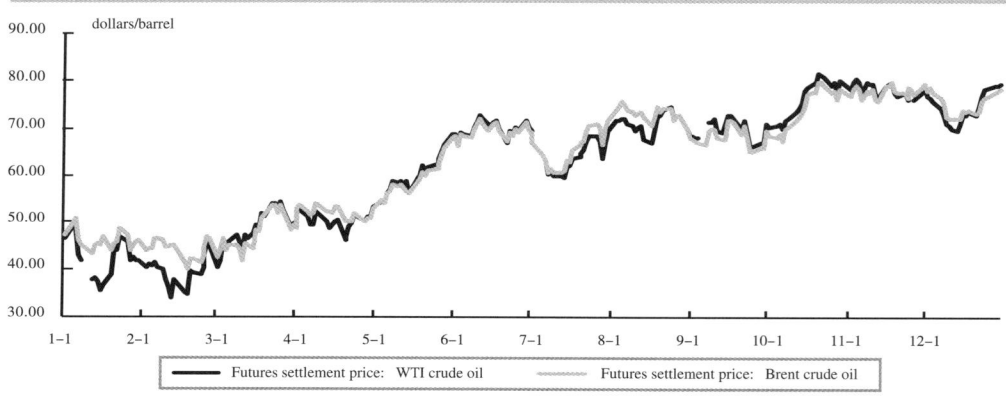

Futures settlement price: WTI crude oil   Futures settlement price: Brent crude oil

Data source: Nymex, IPE

International oil price trends in 2010

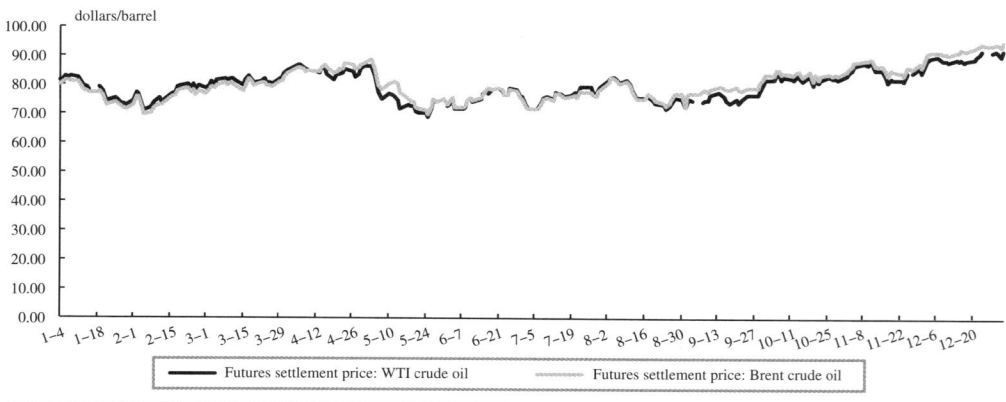

Futures settlement price: WTI crude oil   Futures settlement price: Brent crude oil

Data source: Nymex, IPE

In 2009, the world oil import and export trade volume was 2.61 billion tons, falling by 3.4% compared with that in 2008. Among that, the trade volume of crude oil was 1.89 billion tons, with 0.71 billion tons of oil products.

In 2010, the trade volume of world oil imports and exports was 2.63 billion tons, showing a small-scale growth. Among that, the crude oil trade volume was 1.88 tons, about 70% of the total, and the volume of oil products was 0.76 billion tons.

Import

The USA is the largest oil importer in the world.

In 2009, the oil imports of the USA were 0.57 billion tons, among which crude oil imports were 0.44 billion tons. The volume of imports from Canada and Mexico in all amounted to 0.18 billion tons, accounting for 32.4% of the total imports; imports from the Middle East region further fell to 0.09 billion tons, making up 15.4% of its total imports.

In 2010, the oil imports of the USA were 0.58 billion tons, with crude oil imports being 0.46 billion tons. Canada was still its major source of imports.

In 2009, China's oil imports surpassed that of Japan, becoming the second largest oil importer in the world.

In 2009, China's oil imports were 0.25 billion tons, among which crude oil imports were 0.2 billion tons; its major source of imports was the Middle East, from which the imports were 0.1 billion tons, accounting for 40.7% of its total imports.

In 2010, China still kept its place as the world's second largest oil importer, with the total oil imports for the year being 0.29 billion tons, a year-on-year increase of 16.3%. Among this, crude oil imports were 0.23 billion tons. The Middle East still was the major source of imports, with a total of 0.12 billion tons imported to China.

Japan is the third largest oil importer in the world.

In 2009, Japan's oil imports were 0.21 billion tons, with crude oil imports being 0.18 billion tons; the major source of imports was the Middle East, imports being 0.11 billion tons, accounting for 70.5% of its total imports.

In 2010, Japan's oil imports were 0.23 billion tons, crude oil imports being 0.18 billion tons; the Middle East was still the major source of imports, with 0.12 billion tons from there.

The major importers in Europe are Germany, France, Italy and Spain.

In 2009, the total oil imports in Europe amounted to 0.66 billion tons, with crude oil imports making up 0.51 billion tons; its imports were primarily from the former Soviet Union and the Middle East, with 0.35 billion and 1.1 billion tons respectively, followed by North Africa and West Africa, with 0.08 billion tons and 0.05 billion tons respectively.

In 2010, the total oil imports in Europe were 0.59 billion tons, with crude oil imports being 0.47 billion tons; its major source of imports was the former Soviet Union and the Middle East, with 0.3 billion and 0.12 billion tons respectively.

Export

The Middle East is the world's largest oil importer.

In 2009, total exports in the Middle East accounted for 35.1% of the world total, the total oil exports for the year being 0.91 billion tons, among which crude oil exports were 0.82 billion tons. It was followed by the former Soviet Union, West Africa, Central and South America and North Africa, their total oil exports for the year being 0.45 billion tons, 0.22 billion tons, 0.18 billion tons and 0.14 billion tons respectively.

In 2010, the Middle East was still the world's largest exporter of crude oil, its total oil exports for the year being 0.94 billion tons. It was followed by the former Soviet Union, West Africa, Central and South America and North Africa, their total oil exports being 0.42 billion tons, 0.22 billion tons, 0.18 billion tons and 0.14 billion tons respectively.

World oil trade volumes in 2009

	Imports (million tons)		Exports (million tons)	
	Crude oil	Oil products	Crude oil	Oil products
USA	442.8	122.0	2.2	89.5
Canada	39.1	15.3	96.5	25.7
Mexico	0.5	21.0	63.8	8.0
Middle and South America	25.1	41.3	128.9	54.4
Europe	513.3	152.0	23.1	72.9
Former Soviet Union	0.9	3.2	342.0	105.1
Middle East	7.0	10.5	822.1	91.6
North Africa	18.4	10.0	111.1	25.3
West Africa	#	12.1	212.3	5.3
East and South Africa	21.9	5.7	14.8	0.3
Australia	22.8	17.1	12.8	2.0
China	203.5	49.8	4.7	29.4
India	145.8	10.4	0.1	35.4
Japan	176.5	35.3	—	16.5
Singapore	46.3	79.8	2.3	72.0
Other Asian countries	228.6	127.6	40.2	59.9
Unconfirmed	—	0.9	15.5	20.6
World total	**1892.5**	**714.0**	**1892.5**	**714.0**

Note: # Means less than 0.05

Data source: BP Statistical Review of World Energy, June 2010

World oil trade volume in 2010

	Imports (million tons)		Exports (million tons)	
	Crude oil	Oil products	Crude oil	Oil products
USA	456.1	121.0	1.4	101.7
Canada	28.9	12.7	99.1	29.1
Mexico	0.4	30.1	67.8	8.5
Middle and South America	20.9	56.8	131.2	44.6
Europe	465.1	131.7	19.3	71.8
Former Soviet Union	#	4.8	318.0	103.2
Middle East	11.3	10.1	828.7	107.2
North Africa	12.3	12.0	112.6	29.2
West Africa	0.1	6.9	221.2	7.6
East and South Africa	5.0	7.3	16.2	0.4
Australia	29.0	14.1	16.2	7.6
China	234.6	59.9	2.0	29.4
India	162.0	16.5	0.0	57.2
Japan	184.8	40.9	0.3	14.1
Singapore	39.9	100.1	2.1	65.8
Other Asian countries	225.5	131.7	39.7	80.2
Unconfirmed	—	1.2	—	—
World total	**1875.8**	**757.7**	**1875.8**	**757.7**

Note: # means less than 0.05

Data source: BP Statistical Review of World Energy, June 2011

Coal

In 2009, the economic recession caused a rapid decline in global energy demand. From the beginning of the year till March, coal prices maintained the callback of coal prices in 2008. On 19 March, 2009, the spot price of BJ steam coal in Australia dropped from the highest at 190.95 dollars/ton to 61.5 dollars/ton. However, after coal prices hit bottom in March, and under the influence of such factors as a large-scale increase in coal imports in China, continued dollar depreciation and fluctuating recovery of international oil prices, etc., coal prices began to rebound upwards and continued to go up with some fluctuation.

In 2010, influenced by the demand recovery, dollar depreciation, inflation, adverse weather, etc., national coal prices witnessed a large-scale rise. In early 2010, domestic coal prices rose to the level of the year 2008. In the period from April to May and that from late September to the end of November, domestic coal prices witnessed two rounds of rising tendency separately.

Price trend of BJ steam coal in Australia during the years 2000–2010

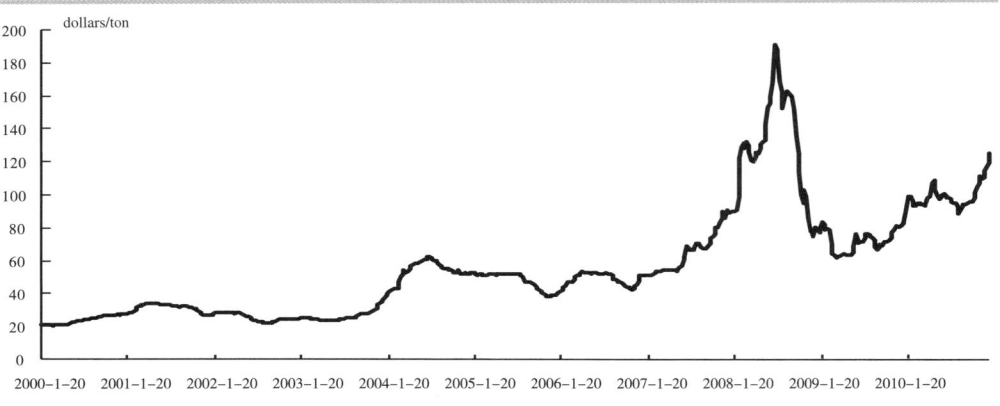

Data source: China Coal Market Network

In 2009, the prices of various kinds of coal in Qinhuangdao Port, China, as a whole, fluctuated upward. In early 2009, as the inventory levels gradually went up, the prices of various kinds of coal in Qinhuangdao Port underwent a fall to different extents. Then with a continued drop in its inventory levels, coal prices saw a rise of different degrees. In late July, due to a continued slow fall in carbon inventory, the prices of coal of every kind correspondingly underwent another round of continuous increase. In November, with the peak of coal use for heating in winter, as well as with a steady economic recovery, the demand for electricity and coal witnessed a rapid growth, hence, coal prices in Qinhuangdao experienced a round of rapid increase.

In 2010, coal prices in Qinhuangdao Port showed a W-pattern trend, i.e. a falling-rising-falling tendency. In early 2010, the great coal demand led to a relatively high price level; as the temperature went up, coal demand and prices began to gradually fall to the lowest level in March of the year; in April coal prices began a low growth until late May, when it stopped increasing and saw a steady period; at the

Coal price trends for China's Qinhuangdao

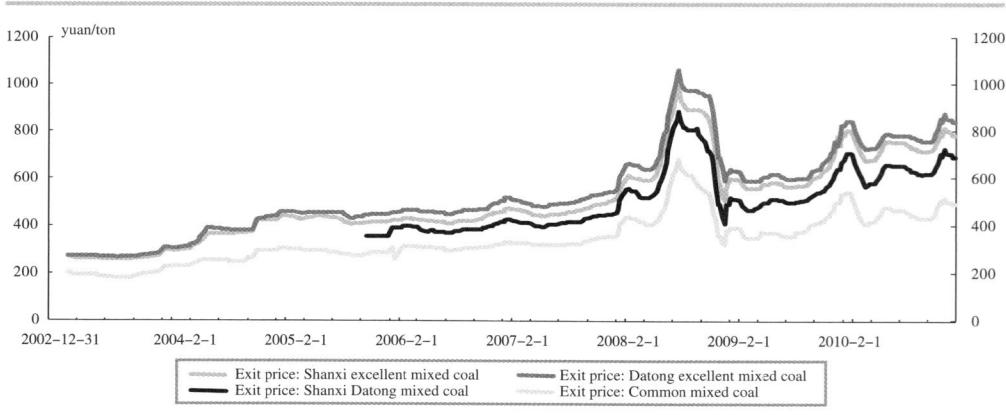

Data source: China Coal Market Network

end of September, the Qinhuangdao Port coal market witnessed another round of recovery; from late October onwards, influenced by a rise in international energy prices and coal storage for the approaching winter, coal prices experienced a rapid growth.

In 2009, the global trade volume of coal shipped was about 0.9 billion tons, with the Asia-Pacific Region boasting the largest coal trade volume the world over, accounting for 50% of the world total. Moreover, an important change in 2009 was that China became for the first time a net importer of coal. This was mainly attributed to the price inversion between the global and the domestic. The domestic coal prices fluctuated at a high level, which greatly encouraged power enterprises to import coal from abroad.

According to the research report issued by the consultancy Clarkson, an independent authority in the field of the international shipping industry, in 2008, the global offshore trade volume of coal was 0.797 billion tons, among which the trade volume of steam coal was 0.578 billion tons and that of coking coal was 0.218 billion tons; in 2009, the global offshore trade volume of coal was 0.795 billion tons, slightly lower than the level in the previous year. However, compared with the past year, the steam coal trade volume saw an increase, up to 0.586 billion tons, and that of coking coal was 0.209 billion tons.

Clarkson predicts that in 2010, the global offshore trade volume of coal will soar to 0.814 billion tons, increasing approximately by 20 million tons over the previous year. Among this, the steam coal trade volume was 0.599 billion tons and that of coking coal rose to 0.215 billion tons.

Natural gas

In 2009, natural gas prices all over the world witnessed a fall to different extents. In Asia, the average import price of liquefied gas in Japan was 9.06 dollars/million British thermal unit (MBtu), dropping by 3.49 dollars/MBtu over the previous year. In the USA, the import price of natural gas declined from 8.85 dollars/MBtu in 2008 to 3.89 dollars/MBtu. In comparison, in the European Union, the CIF price of natural gas fell from 11.56 dollars/MBtu in 2008 down to 8.52 dollars/MBtu; and the natural gas price in Alberta, Canada, also decreased from 7.99 dollars/Mbtu in 2008 to 3.38 dollars/Mbtu.

In 2010, though world natural gas prices went up, the situation where the supply of oil and gas resources exceeded the demand had not been completely turned around. Therefore, the prices remained at a relatively low level. The average import price of liquefied gas in Japan was 10.91 dollars/MBtu; that of the USA was 4.39 dollars/MBtu. For the EU, the CIF price of natural gas was 8.01 dollars/MBtu, and that of Alberta was 3.69 dollars/MBtu. Except for the EU, all the other countries concerned saw a rise in natural gas prices to different degrees.

World natural gas prices

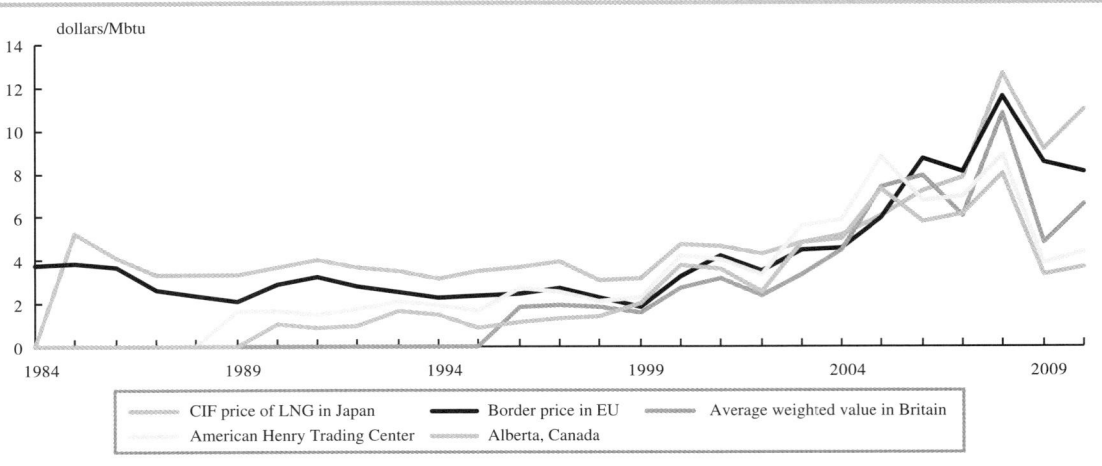

Data source: BP Statistical Review of World Energy, June 2011

In 2009, the world trade volume of natural gas amounted to 876.85 billion m³, increasing by 7.7% over the previous year. Among this, the trade volume of pipeline natural gas was 633.77 billion m³, making up 72.3% of the total; the LNG trade volume was 242.77 billion m³, with a share of 27.7%.

In 2010, natural gas trading all over the world still mainly dealt in pipeline gas, but the role played by LNG became increasingly significant. The total trade volume for the whole year was 975.22 billion m³, a significant growth by 11.3%, compared with the previous year. Among this, the trade volume of pipeline natural gas was 677.59 billion m³, making up 64.5% of the total trade volume; the LNG trade volume saw a substantial increase, up to 297.63 billion m³, with its share in the total soaring to 30.5%.

International trade volumes of natural gas during the years 2002–2010

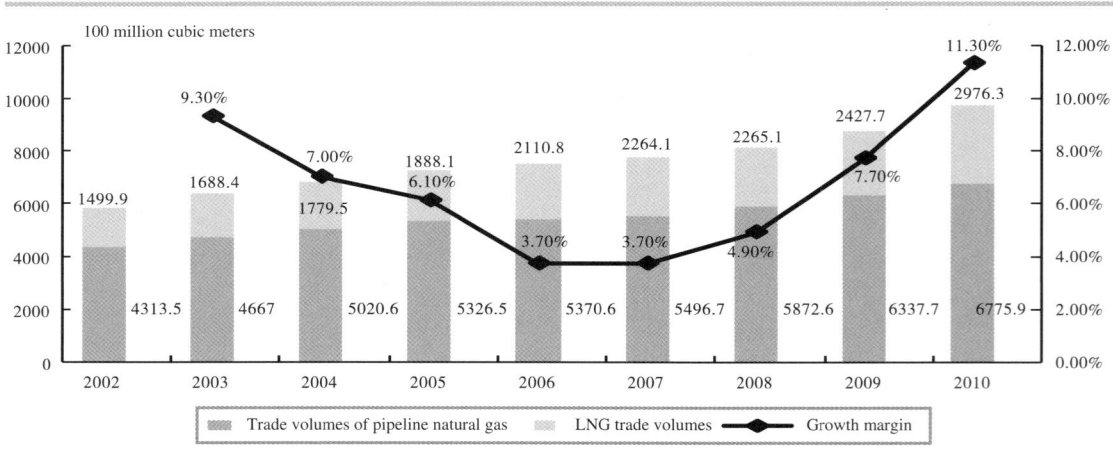

Data source: BP Statistical Review of World Energy, June 2011

Pipeline natural gas

Export

The five largest exporters of pipeline natural gas the world over are Russia, Canada, Norway, Holland and Algeria, in descending order.

In 2009, the sum total exports of these five countries accounted for 70.4% of the global exports. In 2010, this sum total made up 65.6% of the world total exports.

Russia is the world's largest exporter of pipeline gas.

In 2009, the pipeline gas exports in Russia were 176.48 billion m^3, increasing by 14.3% over the previous year, and making up 27.8% of the global total exports. Furthermore, all of its natural gas was exported to European countries. In 2010, its pipeline gas exports were 186.45 billion m^3, rising by 5.7% over the previous year.

Canada is the second largest pipeline gas exporter in the world.

In 2009, its pipeline gas exports were 92.24 billion m^3, falling by 10.6% over the year 2008, and all pipeline gas was exported to the USA; in comparison, the pipeline gas exports in 2010 were 92.40 billion m^3, remaining basically at the same level as the previous year.

Import

In 2009, the USA, Germany, Italy, France and Russia ranked as the top five in the world, whose imports in all accounted for 50.0% of the world total imports. In 2010, the world top five importers of pipeline natural gas were the USA, Germany, Italy, Britain and Ukraine, whose imports in all made up 47.3% of the world total.

The USA is the largest pipeline gas importer in the world.

In 2009, its pipeline gas imports were 104.41 billion m^3, dropping by 10.9% over the previous year, its major source of imports being Canada (92.24 billion m^3) and Mexico (0.79 billion m^3); in 2010, its pipeline gas imports amounted to 93.25 billion m^3, a continued fall of 10.7%.

Germany is the second largest pipeline gas importer in the world, and its major imports are from Russia, Holland and Norway.

Export

In 2009, the major LNG exporters were Qatar, Malaysia, Indonesia, Australia, Algeria, Trinidad and Tobago, and Nigeria. The total LNG exports of these countries as a whole accounted for 76.5% of the world total LNG imports.

In 2010, Qatar still held its first place among the large LNG exporters of the world, with a 53.2% rise in its exports. The second largest LNG exporter was Indonesia, followed by Malaysia and Australia respectively.

LNG trade in 2009

Unit: billion cubic meters

Destination	USA	Trinidad and Tobago	Belgium	Norway	Russia	Oman	Qatar	United Arab Emirates (UAE)	Yemen	Algeria	Egypt	Equatorial Guinea	Libya	Nigeria	Australia	Brunei	Indonesia	Malaysia	Total imports
North America																			
USA	—	6.68	—	0.84	—	—	0.36	—	—	—	4.54	—	—	0.38	—	—	—	—	12.80
Canada	—	0.80	—	—	—	—	0.09	—	—	—	0.08	—	—	—	—	—	—	—	0.98
Mexico	—	0.16	—	0.09	—	—	0.12	—	0.08	—	0.34	—	—	2.69	—	—	0.08	—	3.55
Central and South America																			
Argentina	—	0.80	—	—	—	—	—	—	—	—	0.16	—	—	—	—	—	—	—	0.96
Brazil	—	0.27	—	—	—	—	—	—	—	—	—	—	—	0.08	—	—	—	—	0.35
Chile	—	0.16	—	—	—	—	0.16	—	—	—	—	0.33	—	—	—	—	—	—	0.65
Dominica	—	0.56	—	—	—	—	—	—	—	—	—	—	—	—	—	—	—	—	0.56
Puerto Rico	—	0.76	—	—	—	—	—	—	—	—	—	—	—	—	—	—	—	—	0.76
Europe																			
Belgium	—	0.16	—	0.17	—	—	6.03	—	—	—	0.09	—	—	0.08	—	—	—	—	6.53
France	—	0.72	—	0.44	—	—	0.17	—	—	7.68	1.63	0.08	—	2.35	—	—	—	—	13.07
Greece	—	0.04	—	—	—	—	—	—	—	0.53	0.17	—	—	—	—	—	—	—	0.74
Italy	—	—	—	—	—	—	1.55	—	—	1.27	0.08	—	—	—	—	—	—	—	2.90
Portugal	—	0.40	—	—	—	—	—	0.08	—	0.11	—	0.09	—	2.14	—	—	—	—	2.82
Spain	—	4.18	0.08	1.38	—	1.30	4.98	—	0.09	5.19	4.10	—	0.72	4.99	—	—	—	—	27.01
Turkey	—	0.08	—	—	—	0.08	0.32	—	—	4.20	0.08	—	—	0.94	—	—	—	—	5.71
Britain	—	1.97	—	0.26	—	—	5.75	—	—	1.68	0.51	—	—	—	0.08	—	—	—	10.24

(Continued)

LNG trade in 2009 (*Continued*)

Unit: billion cubic meters

Destination	USA	Trinidad and Tobago	Belgium	Norway	Russia	Oman	Qatar	United Arab Emirates (UAE)	Yemen	Algeria	Egypt	Equatorial Guinea	Libya	Nigeria	Australia	Brunei	Indonesia	Malaysia	Total imports
Middle East																			
Kuwait	—	0.15	0.08	—	0.41	0.08	—	—	—	—	—	—	—	—	0.08	—	—	0.09	**0.89**
Asia-Pacific Region																			
China	—	0.08	0.08	—	0.25	0.09	0.55	—	—	—	0.08	0.08	—	0.08	4.75	—	0.72	0.88	**7.63**
India	—	0.68	—	—	0.67	0.35	8.25	0.17	—	0.16	0.33	0.25	—	0.32	1.12	—	0.08	0.25	**12.62**
Japan	0.86	0.14	—	—	3.69	3.44	10.29	6.75	—	—	0.24	1.70	—	0.77	15.87	8.11	17.25	16.79	**85.90**
Korea	—	0.90	—	—	1.35	6.05	9.28	—	0.25	0.08	0.31	1.52	—	0.23	1.75	0.70	4.10	7.81	**34.33**
Taiwan, China	—	0.08	—	—	0.24	0.16	1.56	—	—	—	0.08	0.67	—	0.93	0.60	—	3.77	3.71	**11.79**
Total exports	**0.86**	**19.74**	**0.24**	**3.17**	**6.61**	**11.54**	**49.44**	**7.01**	**0.42**	**20.90**	**12.82**	**4.72**	**0.72**	**15.99**	**24.24**	**8.81**	**26.00**	**29.53**	**242.77**

Data source: BP Statistical Review of World Energy, June 2010

LNG trade in 2010

Unit billion cubic meters

Destination	USA	Trinidad and Tobago	Peru	Belgium	Norway	Russia	Algeria	Egypt	Equatorial Guinea	Libya	Nigeria	Oman	Qatar	United Arab Emirates (UAE)	Yemen	Australia	Brunei	Indonesia	Malaysia	Total imports
North America																				
USA	—	5.38	0.45	—	0.76	—	—	2.07	—	—	1.18	—	1.29	—	1.10	—	—	—	—	12.23
Canada	—	1.59	0.08	—	0.08	—	—	—	—	—	—	—	0.25	—	—	—	—	—	—	2.00
Mexico	—	—	0.26	—	—	—	—	0.16	—	—	2.23	—	1.02	—	0.18	—	—	1.87	—	5.72
Central and South America																				
Argentina	1.63	—	—	—	—	—	—	—	—	—	—	—	0.15	—	—	—	—	—	—	1.78
Brazil	0.09	0.85	0.16	0.08	—	—	—	—	0.08	—	0.89	—	0.59	0.04	—	—	—	—	—	2.78
Chile	—	0.52	—	—	—	—	0.17	0.55	1.50	—	—	—	0.25	—	0.08	—	—	—	—	3.07
Dominica	—	0.82	—	—	—	—	—	—	—	—	—	—	—	—	—	—	—	—	—	0.82
Puerto Rico	—	0.77	—	—	—	—	—	—	—	—	—	—	—	—	—	—	—	—	—	0.77
Europe																				
Belgium	0.05	0.08	0.08	—	0.09	—	—	0.17	—	—	0.16	—	5.80	—	—	—	—	—	—	6.43
France	—	0.35	—	—	0.51	—	6.27	0.73	—	—	3.57	—	2.43	—	0.08	—	—	—	—	13.94
Greece	—	0.08	—	—	—	—	0.98	0.08	0.03	—	—	—	—	—	—	—	—	—	—	1.17
Italy	—	0.32	—	—	0.16	—	1.61	0.72	0.09	—	—	—	—	—	—	—	—	—	—	9.08
Portugal	—	0.18	—	—	0.05	—	—	—	—	—	2.70	—	0.08	—	—	—	—	—	—	3.01
Spain	0.12	3.32	0.63	0.08	1.64	—	5.08	2.62	—	0.34	7.82	—	5.54	—	0.18	—	—	—	—	27.54
Turkey	—	0.26	—	0.08	0.26	—	3.87	0.27	—	—	1.26	—	—	—	—	—	—	—	—	7.92
Britain	0.18	1.63	—	—	0.94	—	1.25	0.12	—	—	0.40	0.17	13.89	—	0.26	—	—	—	—	18.67

(Continued)

LNG trade in 2010 (*Continued*)

Unit billion cubic meters

Destination	USA	Trinidad and Tobago	Peru	Belg- ium	Norway	Russia	Algeria	Egypt	Equatorial Guinea	Libya	Nigeria	Oman	Qatar	United Arab Emirates (UAE)	Yemen	Australia	Brunei	Indo-nesia	Malaysia	Total imports
Middle East	—	0.33	—	0.09	—	0.09	—	0.33	0.29	—	0.08	0.91	—	0.25	0.09	0.09	—	—	0.24	2.78
Kuwait	—	—	—	—	—	—	—	—	—	—	—	—	0.16	—	—	—	—	—	—	0.16
Asia-Pacific Region																				
China	—	0.07	0.08	0.08	—	0.51	—	0.08	0.08	—	0.17	—	1.61	0.08	0.70	5.21	—	2.45	1.68	12.80
India	—	0.66	—	—	—	—	—	0.09	0.17	—	0.33	—	10.53	—	0.37	—	—	—	—	12.15
Japan	0.85	0.15	—	0.08	—	8.23	0.08	0.57	0.72	—	0.84	3.80	10.15	6.86	0.16	17.66	7.78	17.00	18.55	93.48
Korea	—	0.88	0.08	0.08	0.16	3.90	—	0.98	1.85	—	1.18	6.11	10.16	0.25	2.27	1.33	1.05	7.42	6.39	44.44
Taiwan, China	—	0.51	—	—	0.07	0.67	—	0.17	0.35	—	1.09	0.50	3.75	0.42	—	1.06	—	2.62	3.68	14.90
Total exports	1.64	20.38	1.82	0.57	4.71	13.40	19.31	9.71	5.16	0.34	23.90	11.49	75.75	7.90	5.48	25.36	8.83	31.36	30.54	297.63

Data source: BP Statistical Review of World Energy, June 2011

Import

In 2009, Japan became the world's largest LNG importer, with a total volume of 85.90 billion m³, accounting for 35.4% of the world total imports. Following next were Korea, Spain and France, with a share of 14.1%, 11.1% and 5.4% respectively in the world total LNG imports.

In 2010, Japan was still the world's largest LNG importer, whose imports rose by 8.8%, up to 93.48 billion m³, making up 31.4% of the world total. Next were Korea, Spain and Britain, with their shares in the world total being 14.9%, 9.3% and 6.3% respectively.

Section 2: Trade distribution of major types of energy in China

Coal

In 2009, coal imports in China saw a rapid growth, with a cumulative total of 125.83 million tons, up 211.5% over the previous year. In contrast, coal exports witnessed a continuous decline, with a cumulative total volume of 22.40 billion tons, its decrease margin being 50.7%; for the whole year, net imports were 0.103 billion tons, recording a trade deficit of 8.20 billion dollars, hence China becoming a net importer of coal for the first time.

In 2010, China's coal imports continued to maintain a relatively rapid growth, whose imports recorded a new high, with a cumulative import volume for the year of 164.78 million tons and a year-on-year growth rate of 30.9%. In comparison, its exports continued a falling tendency during recent years, with its cumulative export volume for the year being 19.03 million tons, a year-on-year drop of 15%, a new low since 1991.

Coal import/export in China

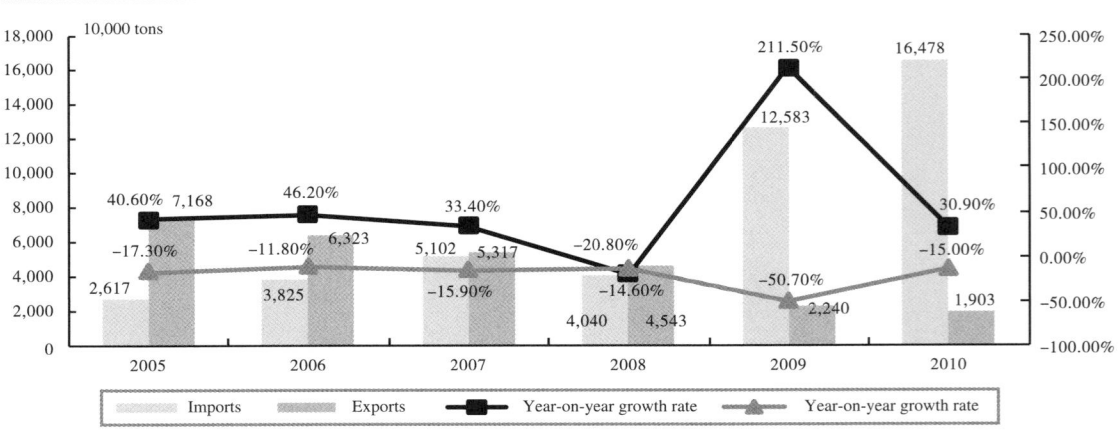

Data source: General Administration of Customs (China)

The reasons for China's becoming a net importer of coal in 2009 are as follows:

First, it is attributed to the overseas/domestic price inversion along with a substantial fall in internal coal prices and shipping fees. For the electric power plants in the coastal cities of southern China, the cost of buying coal from abroad is lower than buying at home, which leads to a soar in coal imports.

Second, our country took the lead in economic recovery, with the increase in domestic coal demand causing coal prices to soar. However, the main coal-production area — Shanxi Province, pioneered in the special renovation of most coal mines, its coal output declined and failed to meet the domestic demand.

Third, our country launched certain policies to restrict export, especially those to cancel export tax rebates, which had a direct impact upon coal import and export.

Import

In 2009, imports of coal of various kinds increased to different degrees. Data from customs shows that Byerlyte and general bituminous coal boasted the largest growth margins, which was mainly due to the influence of a continuous decline in coal imports within countries like Japan and other regions. Besides, as Australian coal sellers took initiatives to expand their business to the Chinese market, imports of coking coal and steam coal from Australia substantially soared. In the period between January and November, the import volumes of anthracite, Byerlyte, steam bituminous coal and other kinds of coal were 31.11 million tons, 30.996 million tons, 32.382 million tons and 15.766 million tons respectively, up 12.494 million tons, 24.846 million tons, 23.002 million tons and 11.752 million tons respectively over the same period of the previous year.

In 2010, the largest imports were those of steam bituminous coal and other kinds of coal. During the above-mentioned months, this figure was up to 82.518 million tons, rising by 34.370 million tons over the same period of the previous year; following next was the import volume of Byerlyte, amounting to 41.816 million tons, with a year-on-year growth of 10.820 million tons; imports of anthracite, however, witnessed a fall of 7.670 million tons, down to 23.440 million tons.

In 2009, the countries from which China exported coal most were Australia, Indonesia, Vietnam and Russia, in descending order.

In 2010, Indonesia became the biggest coal importer for China, the other four countries being Australia, Vietnam, Mongolia and Russia. During the months January to November, the total coal imports from the above-mentioned five countries amounted to 123.120 million tons, accounting for 84% of the total imports. Among this, coal imports from Australia, Vietnam and Russia all underwent a fall. In contrast, coal imports from South Africa, the USA, Colombia, etc., all drastically soared.

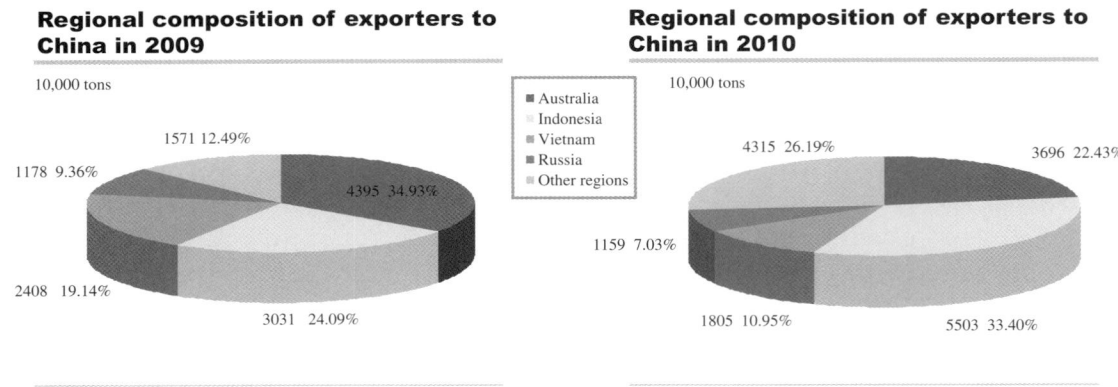

Regional composition of exporters to China in 2009

10,000 tons

1571 12.49%
1178 9.36%
4395 34.93%
2408 19.14%
3031 24.09%

Australia
Indonesia
Vietnam
Russia
Other regions

Regional composition of exporters to China in 2010

10,000 tons

4315 26.19%
3696 22.43%
1159 7.03%
1805 10.95%
5503 33.40%

Data source: General Administration of Customs (China)

Export

In 2009, influenced by such factors as the weakened international market demand, exports of various kinds of coal in China experienced a drop to different degrees.

In terms of absolute value, the decrease margin of general bituminous coal exports was the most significant, falling by 15.348 million tons. This was, for one thing, mainly due to the decline in coal demand from countries to whom China traditionally exported coal, such as Japan and Korea. For another, it was caused by the weakened international market demand, which gave rise to relatively low international coal prices. It meant that profitability from coal exports largely went down, which in turn led to a low degree of motivation among domestic coal exporters.

The largest decrease margin was that of Byerlyte exports. During the period January to November, exports of general bituminous coal dropped by 15.348 million tons, mainly attributed to the reality that Shanxi Province, the major domestic production zone of coking coal, witnessed a fall in its output.

In 2010, there was little obvious change in China's coal exports.

In the two years 2009 and 2010, the top three countries and regions China exported coal to were still Korea, Japan and Taiwan, China.

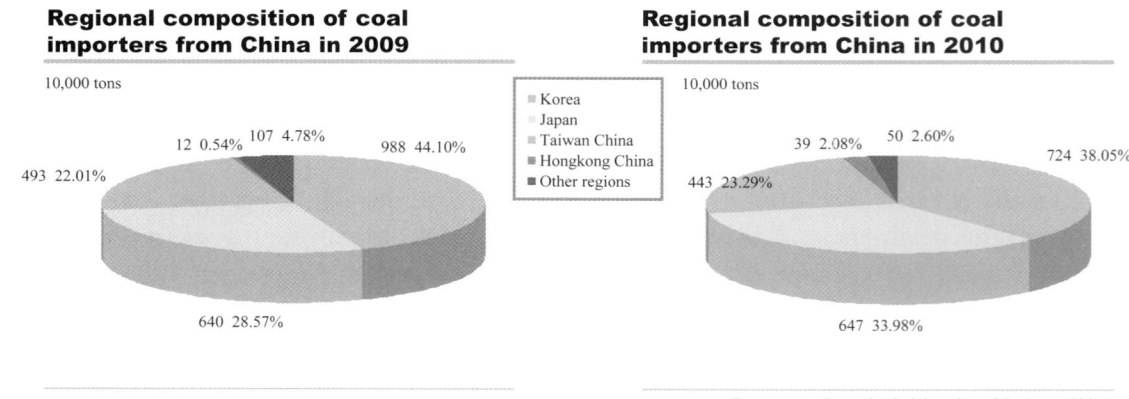

Regional composition of coal importers from China in 2009

10,000 tons

12 0.54%
107 4.78%
988 44.10%
493 22.01%
640 28.57%

Korea
Japan
Taiwan China
Hongkong China
Other regions

Regional composition of coal importers from China in 2010

10,000 tons

39 2.08%
50 2.60%
724 38.05%
443 23.29%
647 33.98%

Data source: General Administration of Customs (China)

Oil

In 2009, China's crude oil imports recorded a new high, up to 0.20 billion tons, a year-on-year rise of 13.9%. The cumulative import turnover amounted to 89.26 billion dollars, with a year-on-year drop of 31.0%; within the same year, the cumulative exports of crude oil were 5.070 million tons, with a year-on-year growth rate of 21.9%; in comparison, its cumulative export turnover was 2.16 billion dollars, with a year-on-year fall of 28.2%.

In 2010, although domestically China saw a large-scale growth in crude oil output, it still failed to catch up with the increase in oil consumption. Under this influence, oil imports witnessed a substantial rise, constantly hitting a record high. Crude oil imports for the whole year amounted to 0.24 billion tons, a year-on-year growth rate of 17.4%.

Crude oil import/export volumes in China from 2005 to 2010

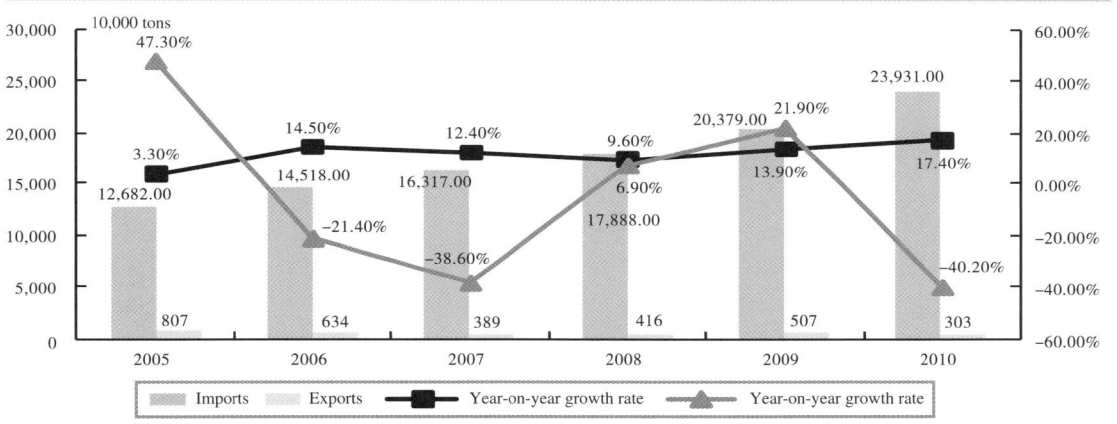

Data source: General Administration of Customs (China)

Crude oil import/export turnovers in China from 2005 to 2010

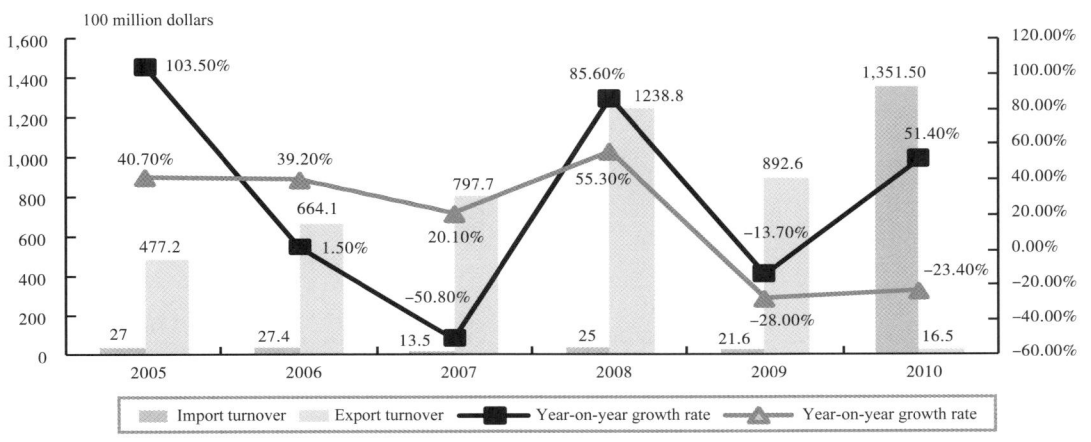

Data source: General Administration of Customs (China)

In 2009, during the first quarter, under the circumstances where China's economic growth slowed down, together with a fall in oil product demand, the monthly crude oil imports also showed a descending trend. Since March, driven by the market expectation that domestic processed oil prices would go up, the import volume of crude oil began to recover. During the last half of the year, pushed by the government's plans to stimulate the economy, the demand for oil products began to recover step by step, along with an accelerated increase in crude oil imports.

In 2010, in the situation where China's crude oil output could not keep up with the increase in consumption, the monthly net imports of oil showed a fluctuating increase, and hit a record high in March, April, June and September separately.

China's crude oil imports in 2009

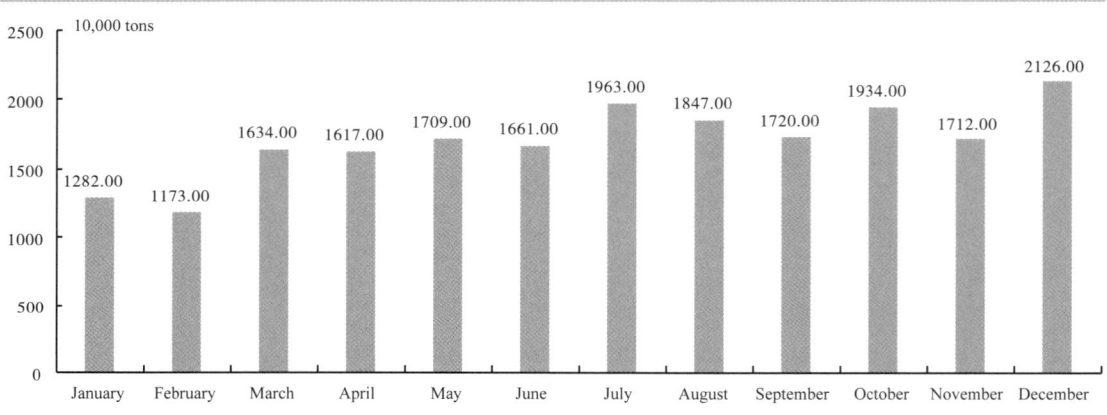

Data source: General Administration of Customs (China)

China's crude oil imports in 2010

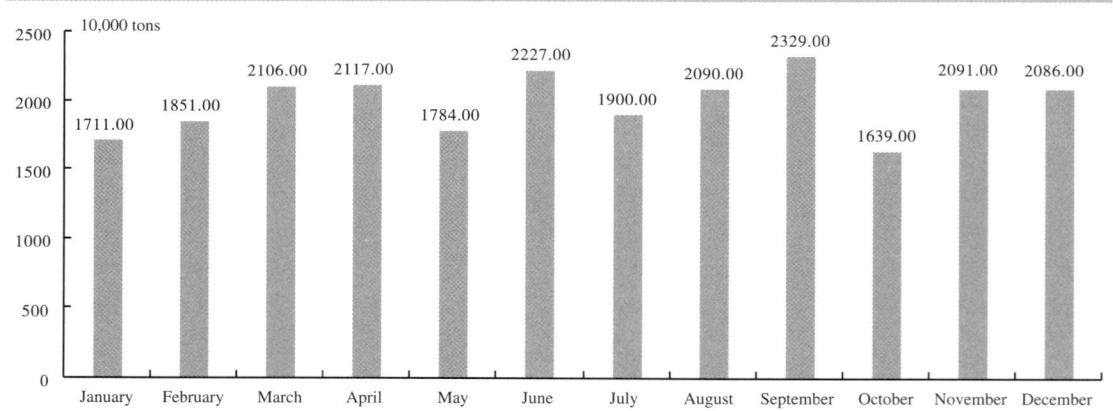

Data source: General Administration of Customs (China)

In terms of exporters to China, the Middle East is the largest, followed by West Africa, with Russia and Central Asia ranking third.

In terms of countries, in 2009, the largest exporting supplier to China was Saudi Arabia with a share of 20.5%, followed by Angola with 15.7%, Iran with a share of 11.3%, Russia 7.5%, Sudan 5.9%, Oman 5.7%, Iraq 3.5%, Kuwait 3.5%, Libya 3.1%, Kazakhstan 2.9%, and Libya with a share of 2.5%. In 2010, the top 10 suppliers that China imported crude oil from were Saudi Arabia, Angola, Iran, Oman, Russia, Sudan, Iraq, Kazakhstan, Kuwait and Brazil.

Distribution of crude oil exporters to China in 2010

Country	Crude oil imports (10,000 tons)	Share
Saudi Arabia	4,464.2	18.7%
The Republic of Angola	3,938.1	16.5%
Iran	2,131.9	8.9%
Oman	1,586.7	6.6%
Russia	1,524.0	6.4%
Sudan	1,259.9	5.3%
Iraq	1,123.8	4.7%
Kazakhstan	1,005.4	4.2%
Kuwait	983.0	4.1%
Brazil	804.7	3.4%
Other countries	5,109.3	21.4%

Data source: General Administration of Customs (China)

Natural gas

Since the year 2009 when China became a net importer of natural gas, its import volume has been growing year by year. This is mainly accounted for by the fact that the growth rate of China's natural gas output fell far behind that of its demand, which means that the domestic supply-demand gap has to be filled through importing.

In 2009, China's natural gas imports amounted to 7.64 billion m³, with a foreign trade dependency ratio (FTDR) of 8%. The major suppliers China imported from were Australia, Malaysia and Indonesia, respectively with a share of 62.2%, 11.5% and 9.4% of the total imports.

Since the end of 2009, China has begun to import natural gas, through pipeline, from Turkmenistan, and become the first Asian country to import both LNG and pipeline gas.

In 2010, China's LNG imports were 9.360 million tons, up 69.2% on a year-on-year basis; it imported 4.4 billion m³ of pipeline gas for the very first time.

China's LNG imports during the years 2006–2009

100 million cubic meters, 10^8 m^3

Legend:
- Malaysia
- Indonesia
- Australia
- Nigeria
- Equatorial Guinea
- Egypt
- Russia
- Algeria
- Qatar
- Oman
- Belgium
- Trinidad & Tobago

Data source: General Administration of Customs (China)

In 2009, China imported 5.532 million tons of LNG, a 65.8% rise over the previous year. Seen on a year-round basis, LNG imports in September hit a record high of 0.789 million tons; in October, LNG imports fell to 0.434 million tons; in November, due to the earlier coming of gas storage for winter, its imports rebounded to 0.588 million tons.

In 2010, China imported 9.360 million tons of LNG, a substantial increase by 69.2% over the previous year, so as to meet the increasing domestic demand for relatively clean fuels.

China's LNG imports (volume by month) in 2009

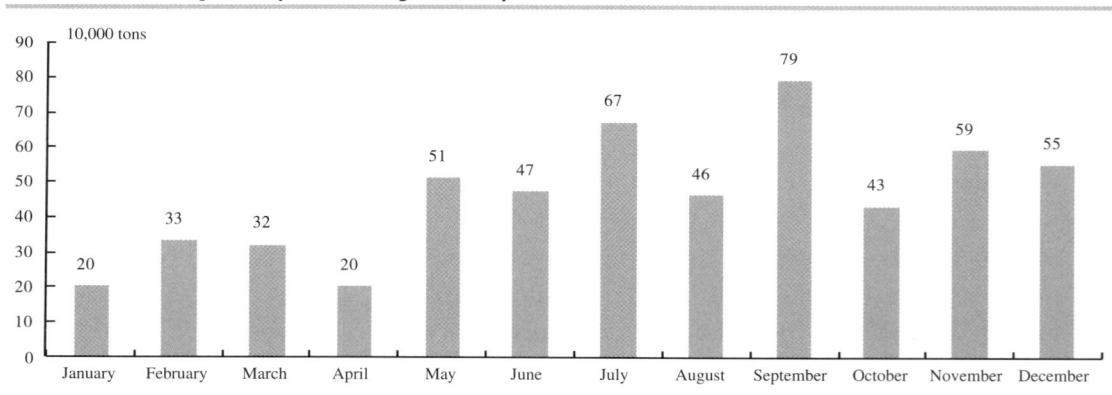

Data source: General Administration of Customs (China)

China's LNG imports (volume by month) in 2010

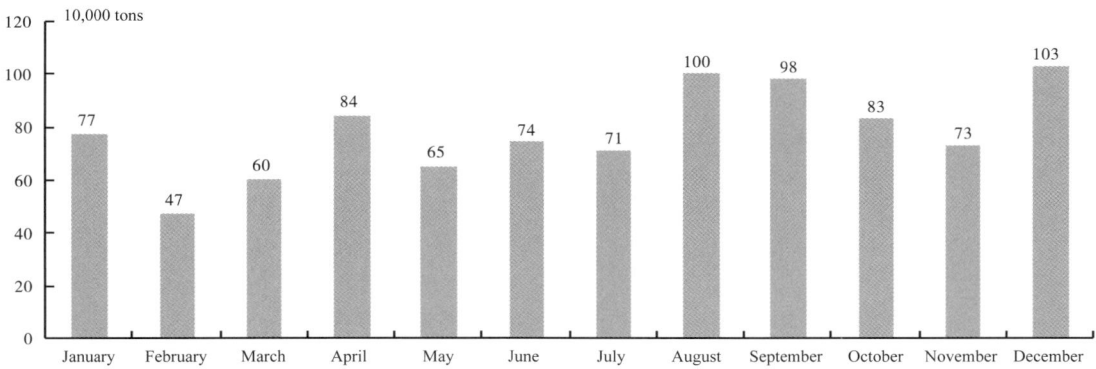

Data source: General Administration of Customs (China)

Chapter 5   Energy Economy

Section 1: Comparative analysis of energy consumption and economies in major countries and China

Since China entered the 21st century, pulled by a continuing economic growth, the energy demand in China has also increasingly gone up, showing a high-speed upward trend. Since 2002, China has ranked second in the world in energy consumption, following closely the USA.

In 2010, the primary energy consumption of China amounted to 2.43 billion tons of oil equivalent (toe), a year-on-year increase of 11.2%, making up 20.3% of the global total of energy consumption. In the USA, primary energy consumption was 2.29 billion toe, a year-on-year rise of 3.7%, and accounting for 9.0% of the world total. China's primary energy consumption surpassed that of the USA, thus ranking first in the world.

In 2010, the top 10 countries in terms of total amount of primary energy consumption were in descending order China, the USA, Russia, India, Japan, Germany, Canada, Korea, Brazil and France.

Primary energy consumption of major countries (Unit: Million toe)											
Country	2000	2001	2002	2003	2004	2005	2006	2007	2008	2009	2010
China	1038.2	1072.5	1140.4	1313.5	1531.3	1691.5	1858.1	1996.8	2079.9	2187.7	2432.2
USA	2313.7	2259.7	2295.5	2302.3	2348.8	2351.2	2332.7	2372.7	2320.2	2204.1	2285.7
Russia	620.4	631.4	633.1	649.9	657.8	657.4	675.3	685.8	691.0	654.7	690.9
India	295.8	297.4	308.7	317.2	345.5	364.0	381.4	414.5	444.6	480.0	524.2
Japan	514.1	512.9	510.3	511.0	522.1	527.2	528.3	523.6	516.2	473.0	500.9
Germany	332.3	338.8	334.0	337.1	337.3	333.2	339.5	324.2	326.8	307.4	319.5
Canada	302.3	298.2	303.1	312.3	315.2	325.3	323.6	329.0	326.6	312.5	316.7
Korea	188.9	193.5	202.4	209.0	213.2	220.6	222.7	231.3	235.3	236.7	255.0
Brazil	185.2	182.9	187.1	192.5	199.9	207.2	212.7	225.4	235.1	234.1	253.9
France	254.2	258.4	255.4	259.3	263.6	261.2	259.2	256.7	257.8	244.0	252.4
Britain	224.1	226.7	221.9	225.6	227.4	228.3	225.6	218.4	214.9	203.6	209.1
Italy	176.5	177.5	176.3	182.3	185.8	186.2	185.4	182.4	180.7	168.3	172.0
South Africa	101.2	100.1	103.4	109.5	115.7	113.5	115.4	118.0	116.3	118.8	120.9
Australia	31.8	33.5	33.2	34.6	34.0	34.8	34.1	33.4	34.1	33.1	33.3

Data source: BP Statistical Review of World Energy, June 2011

Although China's primary energy consumption continues to increase, yet its per unit GDP energy consumption is always showing a descending trend, from 81,100 toe/100 million dollars down to 46,300 toe/100 million dollars within the same year in 2008. This declining tendency has been particularly obvious since 2004. This indicates that in recent years our country has put great emphasis on energy saving and emission reduction, and has achieved noteworthy results.

On the other hand, even under these circumstances, the energy consumption intensity of China still remains at a relatively high level when compared with that of other countries. In 2010, China's per unit GDP energy consumption was thrice that of the USA, and five times that of Japan.

Per unit GDP energy consumption of major countries (Unit: 10,000 toe/100 million dollars)									
Country	2000	2001	2002	2003	2004	2005	2006	2007	2008
USA	2.35	2.22	2.18	2.09	2	1.88	1.76	1.71	1.62
Japan	1.1	1.25	1.3	1.2	1.13	1.15	1.18	1.17	1.03
Britain	1.55	1.57	1.4	1.23	1.05	1.01	0.94	0.78	0.8
Germany	1.73	1.77	1.63	1.36	1.2	1.16	1.13	0.94	0.85
France	1.91	1.92	1.75	1.44	1.27	1.22	1.15	0.99	0.9
Italy	1.59	1.57	1.43	1.19	1.06	1.04	0.98	0.85	0.77
Canada	4.15	4.15	4.1	3.59	3.16	2.85	2.52	2.3	2.36
Australia	2.66	2.86	2.66	2.07	1.76	1.6	1.61	1.34	1.17
Russia	24.07	20.25	18.39	14.83	10.94	8.41	6.86	5.29	4.26
China	8.11	7.6	7.28	7.46	7.38	7.04	6.42	5.67	4.63
India	6.29	6.13	6.11	5.19	4.99	4.48	4.19	3.6	3.56
Brazil	3.04	3.53	3.96	3.63	3.13	2.43	1.87	1.64	1.41
South Africa	8.12	9	9.96	7	5.68	4.95	4.82	4.58	4.78

Data source: http://stats.unctad.org/, World Bank, BP Statistical Review of World Energy 2009

In 2009, in terms of GDP energy consumption in China, the top three largest provinces/regions were Ningxia at 3.454 tons of standard coal/10,000 yuan, Qinghai at 2.689 tons of standard coal/10,000 yuan, and Shanxi at 2.364 tons of standard coal/10,000 yuan. By contrast, the top three in terms of the lowest GDP energy consumption were Beijing at 0.606 tons of standard coal/10,000 yuan, Guangdong at 0.684 tons of standard coal/10,000 yuan, and Shanghai at 0.727 tons of standard coal/10,000 yuan. In 2009, as for per unit GDP energy consumption, the top three in terms of decreasing margin were Gansu with −6.97%, Inner Mongolia with −6.91% and Qinghai with −6.46%.

In 2010, each province in China faced a continuous descending trend in GDP energy consumption. In terms of GDP energy consumption, the top three were Ningxia at 3.308 tons of standard coal/10,000 yuan, Qinghai at 2.550 tons of standard coal/10,000 yuan and Guizhou at 2.248 tons of standard coal/10,000 yuan. By contrast, the lowest three were Beijing at 0.528 tons of standard coal/10,000 yuan, Guangdong at 0.664 tons of standard coal/10,000 yuan and Shanghai at 0.712 tons of standard coal/10,000 yuan.

Per unit GDP energy consumption in various provinces, autonomous regions and municipalities in 2009				
	Per unit GDP energy consumption		Per unit industrial added value energy consumption	
Regions	Index value (tons of standard coal/10,000 yuan)	Up or down (±%)	Index value (tons of standard coal/10,000 yuan)	Up or down (±%)
Beijing	0.606	−5.76	0.909	−12.30
Tianjin	0.836	−6.03	0.911	−13.54
Hebei	1.640	−5.02	2.999	−9.54
Shanxi	2.364	−5.73	4.550	−8.81
Inner Mongolia	2.009	−6.91	3.557	−15.10
Liaoning	1.439	−5.08	2.257	−6.95
Jilin	1.209	−6.19	1.621	−8.19
Heilongjiang	1.214	−5.85	1.382	−9.64
Shanghai	0.727	−6.17	0.957	−5.00
Jiangsu	0.761	−5.17	1.107	−10.17
Zhejiang	0.741	−5.41	1.123	−4.96
Anhui	1.017	−5.39	2.100	−11.13
Fujian	0.811	−3.81	1.150	−2.70
Jiangxi	0.880	−4.54	1.674	−10.13
Shandong	1.072	−5.46	1.543	−9.20
Henan	1.156	−6.16	2.708	−11.56
Hubei	1.230	−5.97	2.350	−12.27
Hunan	1.202	−5.10	1.570	−13.68
Guangdong	0.684	−4.27	0.809	−6.94
Guangxi	1.057	−4.43	2.235	−6.68
Hainan	0.850	−2.81	2.613	−4.53
Chongqing	1.181	−5.50	1.854	−11.95
Sichuan	1.338	−5.83	2.249	−9.18
Guizhou	2.348	−4.12	4.320	−0.03
Yunnan	1.495	−4.60	2.739	−3.78
Shaanxi	1.172	−4.56	1.367	−5.82
Gansu	1.864	−6.97	3.530	−12.84
Qinghai	2.689	−6.46	2.936	−9.46
Ningxia	3.454	−6.26	6.509	−8.71
Xinjiang	1.934	−1.53	3.095	−1.72

Data: NBS

	Energy-saving accomplishment in various regions during the Eleventh Five-Year period			
	2005		2010	
Regions	Per unit GDP energy consumption (tons of standard coal/10,000 yuan)	Planned fall during the Eleventh Five-Year period (%)	Per unit GDP energy consumption (tons of standard coal/10,000 yuan)	Fall by (%) over 2005
Beijing	0.792	−20	0.582	−26.59
Tianjin	1.046	−20	0.826	−21.00
Hebei	1.981	−20	1.583	−20.11
Shanxi	2.890	−22	2.235	−22.66
Inner Mongolia	2.475	−22	1.915	−22.62
Liaoning	1.726	−20	1.380	−20.01
Jilin	1.468	−22	1.145	−22.04
Heilongjiang	1.460	−20	1.156	−20.79
Shanghai	0.889	−20	0.712	−20.00
Jiangsu	0.920	−20	0.734	−20.45
Zhejiang	0.897	−20	0.717	−20.01
Anhui	1.216	−20	0.969	−20.36
Fujian	0.937	−16	0.783	−16.45
Jiangxi	1.057	−20	0.845	−20.04
Shandong	1.316	−22	1.025	−22.09
Henan	1.396	−20	1.115	−20.12
Hubei	1.510	−20	1.183	−21.67
Hunan	1.472	−20	1.170	−20.43
Guangdong	0.794	−16	0.664	−16.42
Guangxi	1.222	−15	1.036	−15.22
Hainan	0.920	−12	0.808	−12.14
Chongqing	1.425	−20	1.127	−20.95
Sichuan	1.600	−20	1.275	−20.31
Guizhou	2.813	−20	2.248	−20.06
Yunnan	1.740	−17	1.438	−17.41
Tibet	1.450	−12	1.276	−12.00
Shaanxi	1.416	−20	1.129	−20.25
Gansu	2.260	−20	1.801	−20.26
Qinghai	3.074	−17	2.550	−17.04
Ningxia	4.140	−20	3.308	−20.09
Xinjiang	Assessed otherwise			

Note: Data about the Tibet Autonomous Region is provided by the government of the Tibet Autonomous Regions.

Data: NBS

Section 2: Energy economic pattern and energy strategy in China

Currently, nearly 90% of world primary energy consumption relies on oil, natural gas and coal, the three major types of fossil energy. Generally, about 70% of China's primary energy supply relies on coal, which severely pollutes the environment and the use of which emits a large amount of soot and SO_2. The over-reliance on coal has severely polluted the environment, thus the urgent call for adjustment and optimization of the energy structure.

Natural gas is the kind of fossil energy that causes least pollution to the environment and boasts a higher heating value (hv) than coal and oil. China plays an important role in saving energy and reducing emissions, which will certainly increase the share of natural gas consumption. This is also the background against which natural gas consumption in China has soared recently.

Index	The Eleventh Five-Year Plan	The Twelfth Five-Year Plan	2011
GDP growth rate (%)	Planned annual average growth: 7.5% Actual annual average growth: 11.2%	Annual average growth of 7%	8%
Decline in per unit GDP energy consumption (%)	Planned index: around 20% Actual performance: 19.1%	16%	3.5% Compared to the previous year
Decline in per unit GDP CO_2 emission (%)	Planning the requirement, and performing well in controlling the emissions of greenhouse gases	17%	—
Decline in total emission of major pollutants (%)	Planned index: 10% Actual performance: chemical oxygen demand (COD) and SO_2 emission falling by 12.45% and 14.29% respectively	COD and SO_2 emissions both falling by 8%: Emissions of ammonia nitrogen and nitrogen oxides both falling by 10%	Emissions of SO_2, COD, ammonia nitrogen and nitrogen oxides all falling by 1.5% compared to the previous year

According to the Outline of the Twelfth Five-Year Plan for National Economic and Social Development (Draft), during the Twelfth Five-Year period, the share of non-fossil energy in primary energy consumption is to rise to 11.4%; per unit GDP energy consumption and CO_2 emission are to be reduced by 16% and 17% respectively; the total emission of major pollutants is to be reduced by 8–10%.

In 2009, during climate talks, China made a commitment to the international community that by 2020, its share of non-fossil energy consumption in the total primary energy consumption would increase to around 15%, and the carbon intensity would decline by 40–45% compared with that in the year 2005.

It is our country's long-term strategy to reasonably control total energy consumption. Since entering into this century, energy demand in China has seen a super-speed growth. Besides, the total production of fossil energy, including coal, oil and electricity, has always been significantly striding forward, quickly making China the second largest energy-consuming country in the world. Therefore, it is our country's long-term strategic choice in the future to control total energy consumption and build an energy-saving nation.

The Twelfth Five-Year energy plan highlights the following six key points:

First, optimize the energy structure. We shall constantly increase the proportion of clean energy such as hydropower, nuclear power, wind power and solar energy; we advocate the conception of consuming environment-friendly and energy-saving forms of energy, accelerate the development of heat-and-power cogeneration, improve the facilities of urban gas pipelines and reasonably utilize renewable energy to increase, step by step, the proportion of non-coal energy in new energy consumption.

Second, adjust the energy industrial pattern. We shall continue to enhance the exploitation and development and comprehensive utilization of traditional energy resources and emerging energy resources. Priority should be given to impelling the development and construction of large-scale energy bases, along with supporting thoroughfare construction for energy transmission. We shall meanwhile promote the restructuring and merging between energy industries, as well as energy industry with other related industries, so as to achieve an intensive and efficient development of the energy industry and relevant industries.

Third, promote energy technology innovation. We shall promote a vigorous development of equipment technology for emerging forms of energy such as wind energy, solar energy, biomass energy and clean coal utilization, nuclear energy, intelligent grid, new-energy cars and distributed energy; gradually export overseas our advanced energy technology, equipment and products; develop a new energy economy befitting China; and realize a great leap from a country with a large amount of energy to a country strong in energy exploitation, development and utilization.

Fourth, improve the macro-control system of energy. We shall expand scientific and rational development and utilization of energy and the macro-control system of energy, lay great emphasis on the constraints of the ecological environment on energy development and utilization, and constantly enhance our capacity to address global climate change. We shall also improve the system of energy strategic reserves and emergency support, and strengthen our comprehensive national regulatory power over energy. Meanwhile, we shall constantly reinforce the construction of the energy infrastructure and public service system, and raise the level of energy "public welfare".

Fifth, intensify energy system reform. We shall launch, in a planned and systematic way, reforms in the energy price system, fiscal and taxation system, resources, circulation system, etc. Meanwhile, we shall proactively cultivate a diversified market entity, and form a modern energy market system that is uniform, open, competitive and orderly.

Sixth, further establish the system of policy standards on sustainable energy development. We shall impel the execution of policies and construction of standards that are beneficial to a healthy development of the energy industry. We shall effectively relieve pressure on energy safety and environmental protection in the near future, and gradually form a new system of sustainable energy development in the middle and long run, to pursue the everlasting energy development.

Key construction projects

Construction of five large national energy bases: The total resource reserves of coal, oil and gas, and hydropower in Shanxi, the Ordos Basin, eastern Inner Mongolia, the southwest region and Xinjiang as a whole make up over 70% of the total amount nationally. From the beginning of the Twelfth Five-Year period onwards, our country will make these five energy-intensive regions into the national strategic energy bases that prop up our long-term national economic development.

Hydropower: Priority shall be given to strengthening the development of key river basins, promoting the construction of large-scale hydropower bases off the upper reach of the Yellow River, Jinshajiang River, Yalong River, Dadu River, Nujiang River, and Langcang River, etc., and completing the building of such large hydropower stations as Xiangjiaba and Xiluodu. We shall rationally lay out the pumped storage power stations and develop small-sized hydropower stations according to local conditions.

Nuclear power: The construction of coastal nuclear power stations shall be arranged and inland nuclear power projects shall be steadily promoted. We shall launch, in an orderly way, the second phases of the Hongyanhe project, Sanmen project and Haiyang project, etc. In addition, we shall launch, at the right time, the first phases of the Taohuajiang project.

Wind power: We shall promote, in an orderly manner, the construction of large-sized wind power bases, with key attention on the 10 million kW wind power bases in Inner Mongolia, Gansu, Xinjiang, Hebei, Jiangsu, Shandong, Jilin and the northeastern regions. Meanwhile, we shall accelerate the development of offshore wind power.

Solar energy: We shall continue to promote the utilization of solar-energy heaters, reinforce support for research and development (R&D) of solar power technology. The state-level R&D and testing center of solar energy will be established, and fiscal and enterprise investment in R&D shall be enlarged. Meanwhile, in the regions abundant in solar energy and those in possession of desert and barren land, a batch of large-scale photovoltaic grid power plants shall be constructed. In suitable regions within Inner Mongolia, Gansu, Qinghai, Xinjiang and Tibet, pilot programs developing thermal-power generation via solar energy shall be launched.

Chapter 6   Energy and the Environment

Section 1: World major indexes of energy and the environment

The increasingly severe global warming is caused, to a large extent, by human activities which lead to the emission of four long-lived greenhouse gases, i.e. carbon dioxide (CO_2), methane (CH_4), nitrous oxide (N_2O) and halocarbon (a group of fluoride-bearing, chlorine-bearing or bromine-bearing gases). When their emissions exceed their cleanup, the concentration of greenhouse gases in the atmosphere will increase, thus causing global warming.

Among the emissions of greenhouse gases all over the world, CO_2 emissions make up the largest share, the increase mainly due to the burning of fossil fules.

In 2009, influenced by the economic crisis and overall GDP decline the world over, global CO_2 emissions saw a fall, yet just 1.8% down compared with the previous year. On the other hand, according to the data released by the Global Carbon Project, CO_2 concentration in the atmosphere was still increasing, and its global average concentration by the end of the year hit 387.2 ppm. Before the industrial revolution, CO_2 concentration was 280 ppm, increasing annually by 2–3 ppm.

In 2010, with a substantial increase in energy demand in various forms, CO_2 emissions also witnessed a rapid growth, up 5.8%, the largest margin since 1969.

World CO_2 emissions (unit: million tons)

2000	25576.9	2.6%
2001	25800.8	0.9%
2002	26301.3	1.9%
2003	27508.7	4.6%
2004	28875.2	5.0%
2005	29826.1	3.3%
2006	30667.6	2.8%
2007	31641.2	3.2%
2008	31915.9	0.9%
2009	31338.8	−1.8%
2010	33158.4	5.8%

Data source: BP Statistical Review of World Energy, June 2010

In 2009, although total CO_2 emissions fell by 1.8%, yet the world total emissions from fossil fuels registered a record of 30.8 billion tons, second only to emissions in the year 2008.

According to the report issued by *Nature Geoscience*, in 2009, the world total CO_2 emissions underwent a slight decline, with a significant regional transfer incurred. Since it was mainly the developed countries that fell victim to the global financial crisis, their emission reduction margins were the most significant: the USA by 6.9%, Britain by 8.6%, Germany by 7%, Japan by 11.8%, Russia by 8.4% and Australia by 0.4%. However, for countries with emerging markets, their emissions witnessed a substantial rise: China by 6.2%, India by 8% and Korea by 1.4%.

In 2009, in the Copenhagen Accord, every country called for a large-scale reduction in global carbon emission in a scientific way so as to cap the global temperature increase at 2°C. According to the research launched by the Intergovernmental Panel on Climate Change (IPCC), the increase in global temperature by 2°C is regarded as a crucial point, which if surpassed, will have a large-scale and irreversible environmental impact on the Earth.

According to the requirements of the Copenhagen Accord, the countries each submitted to the Secretariat of the United Nations Framework Convention on Climate Change their respective greenhouse gas emission reduction commitments by 2010.

Industrial countries	
Country	Emission reduction commitments and base year specified in the Copenhagen Accord
Australia	5–25% (2000)
Canada	17% (2005)
Croatia	5% (before entrance to)
EU	20–30% (1990)
Japan	25% (1990)
Kazakhstan	15% (1992)
New Zealand	10–20% (1990)
Norway	30–40% (1990)
Russia	15–25% (1990)
USA	17% (2005)

Developing countries

Country	Emission reduction commitments and base year specified in the Copenhagen Accord
Brazil	36–39% (BAU)
China	40–45% carbon intensity (2005)
India	20–25% carbon intensity (2005)
Indonesia	26% (BAU)
Israel	20% (BAU)
Maldives	100%
Marshall Islands	40% (2009)
Moldova	25% (1990)
Singapore	16% (BAU)
South Africa	34% (BAU)
Korea	30% (BAU)

Note: "BAU" means "business as usual" pattern

Section 2: Overall condition of China's environment in relation to energy

Investment in environmental pollution control

The Chinese government pays great attention to environmental protection: total investment in environmental pollution control has witnessed a year-on-year increase, with a steady growth in its share of GDP. In 2009, the national total investment in environmental pollution control hit 452,530 million yuan, rising by 3.5 times over the year 2000, with its share of GDP being 1.35%. In 2010, the national total investment in environmental pollution control rose to 665,420 million yuan, with its share of GDP increasing to 1.65%.

National investment in environmental pollution control over the years 2000–2009

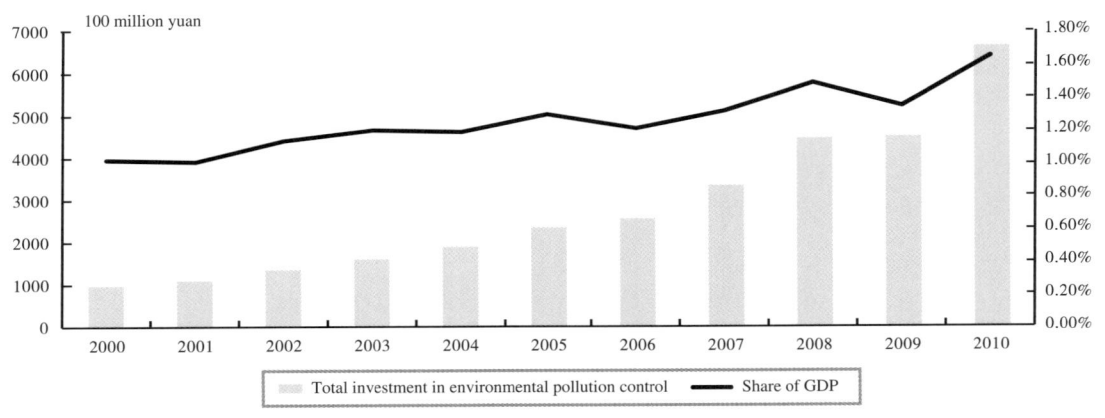

Data: NBS

Primary objective of pollution control

As specified in the Outline of the Eleventh Five-Year Plan for National Economic and Social Development of the People's Republic of China, China put forward the restrictive indexes in the Eleventh Five-Year period as follows: per unit GDP energy consumption is to be reduced by about 20%; total discharge of major pollutants is to be reduced by 10%. By 2010, two restrictive indexes relevant to pollution reduction, chemical oxygen demand (COD) and SO_2 emissions, saw a drop compared to the year 2005, that is, respectively from 14.142 million tons to 12.728 million tons, and from 25.494 million tons to 22.944 million tons.

Major pollutant indicators

In 2007, there appeared a turning point in major pollutant indicators: a successful reduction in COD and SO_2 emissions.

In 2009, China's COD totaled 12.775 million tons, falling by 3.27% over the previous year, and SO_2 emissions amounted to 22.144 million tons, down 4.60% over the year 2008, hence a continuous positive trend of reduction. Compared with 2005, in 2009, the total amount of COD and SO_2 emissions declined by 9.66% and 13.14%, indicating that the reduction margin of SO_2 emissions was well above the reduction target in the Eleventh Five-Year period. COD amounted to 12.381 million tons, falling by 3.09% over the previous year, and SO_2 emissions hit 21.851 million tons, a 1.32% drop over the year 2008. Compared with the year 2005, the total amount of COD and SO_2 emissions fell by 12.45% and 14.29% respectively, both exceeding the 10% emission-reduction target.

Exhaust gases

In 2009, total SO_2 emissions were 22.144 million tons, soot emissions were 8.472 million tons, and industrial dust emissions were 5.236 million tons, dropping by 4.6%, 6.0% and 11.7% respectively over the previous year.

SO₂ emissions in exhaust gases

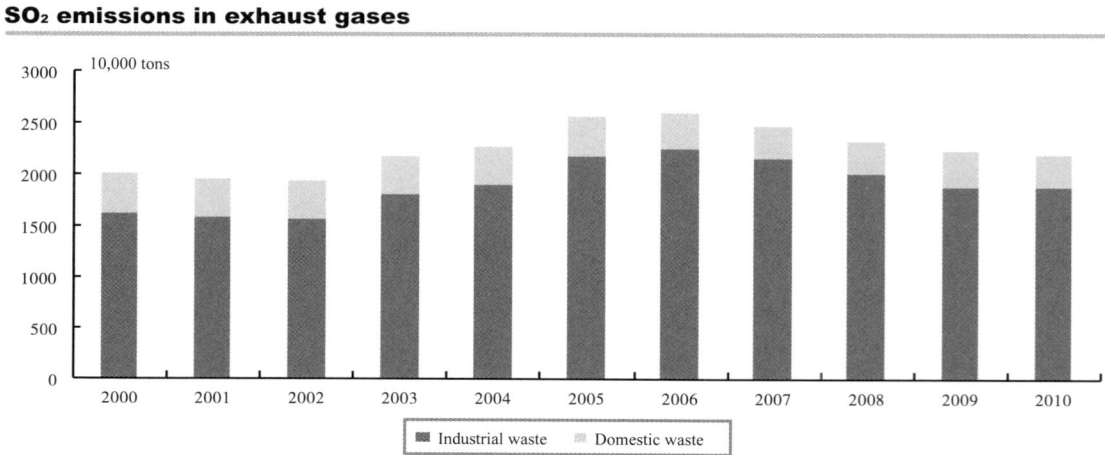

Data: NBS

Soot emissions in exhaust gases

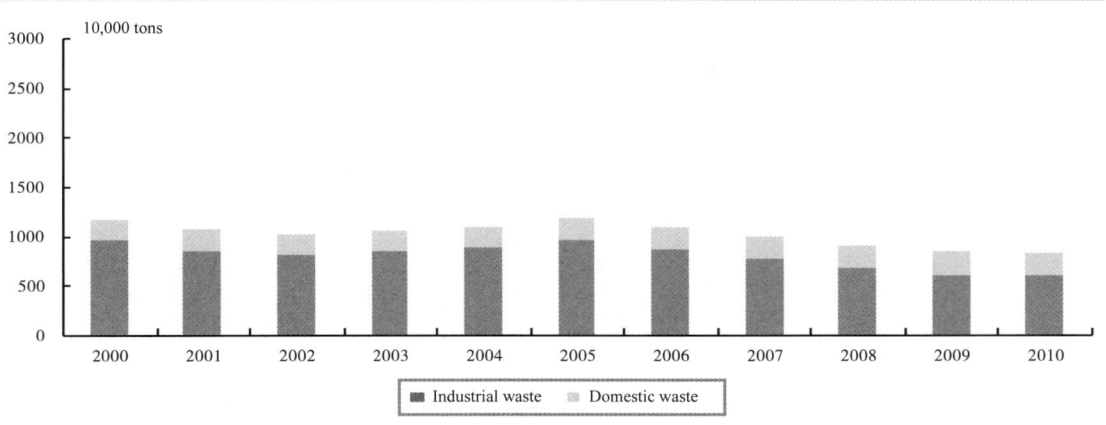

Data: NBS

Industrial dust emissions from pollutants in exhaust gases

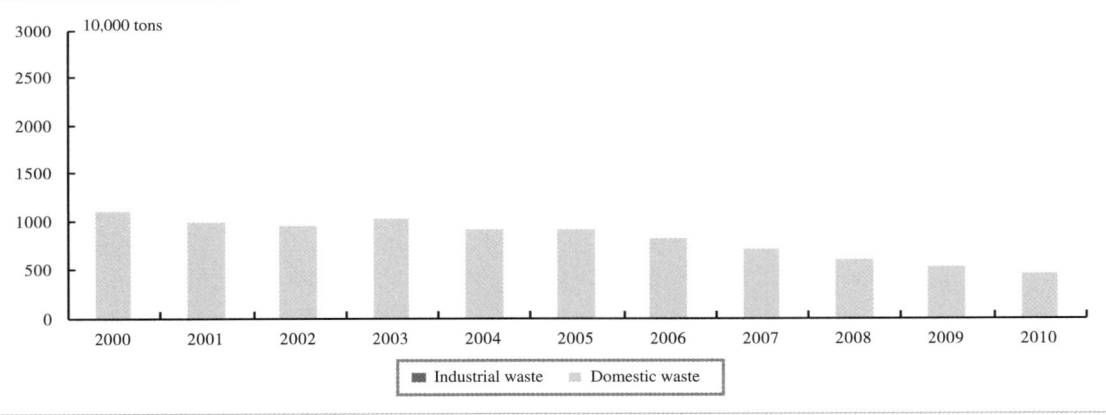

Data: NBS

Wastewater

In 2009, the national wastewater discharge totaled 58.92 billion tons, 3.0% up over the previous year; COD hit 12.775 million tons, a 3.3% drop over the previous year; ammonia nitrogen emissions amounted to 1.226 million tons, falling by 3.5% over the year 2008.

In 2010, the national total wastewater discharge hit 61.73 billion tons, rising by 4.7% over the previous year; that of COD was 12.381 million tons, dropping by 3.1% over the previous year; that of ammonia nitrogen emissions was 1.203 million tons, falling by 1.9% over the year 2008.

National discharge of wastewater and pollutants over the years 2000–2010 (wastewater discharge)

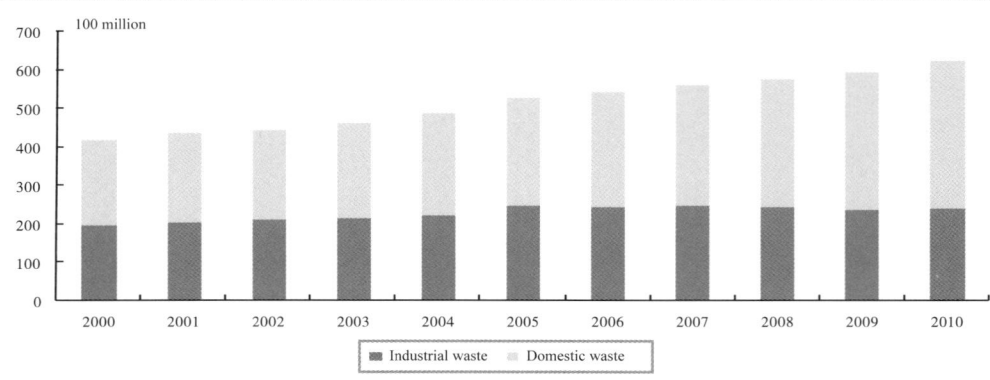

Data: NBS

National discharge of wastewater and pollutants over the years 2000–2010 (COD)

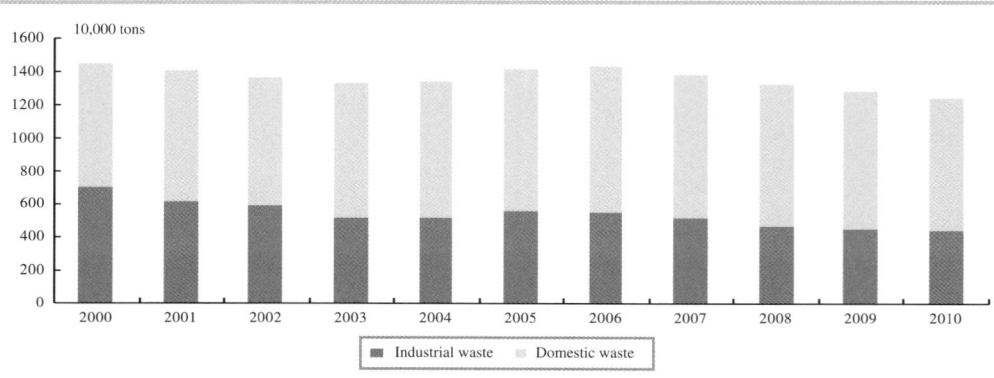

Data: NBS

National discharge of wastewater and pollutants over the years 2000–2010 (ammonia nitrogen discharge)

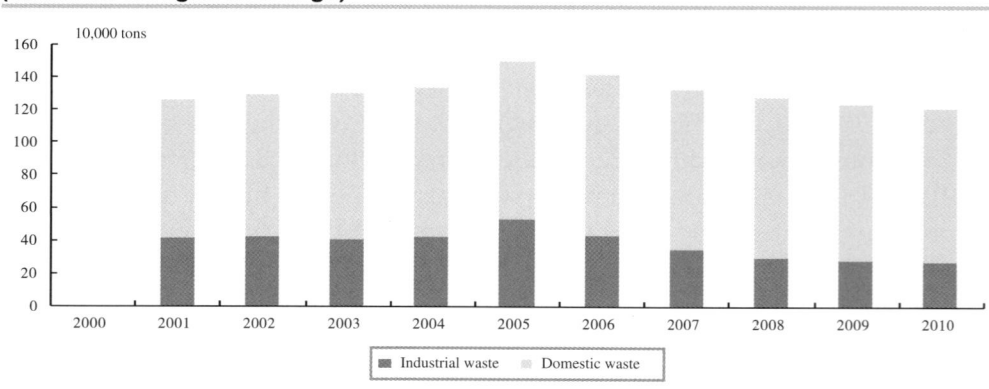

Data: NBS

Industrial solid waste

In 2009, the national industrial solid waste generation totaled 2039.43 million tons, 7.3% up over the previous year, and the total discharge emissions were 7.105 million tons, down 9.1% over the previous year.

In 2010, the national industrial solid waste generation amounted to 2409.44 million tons, rising by 18.1% over the previous year, with its total discharge being 4.982 million tons, dropping by 29.9%.

Generation, discharge and comprehensive utilization of industrial solid waste nationwide (2002–2010)

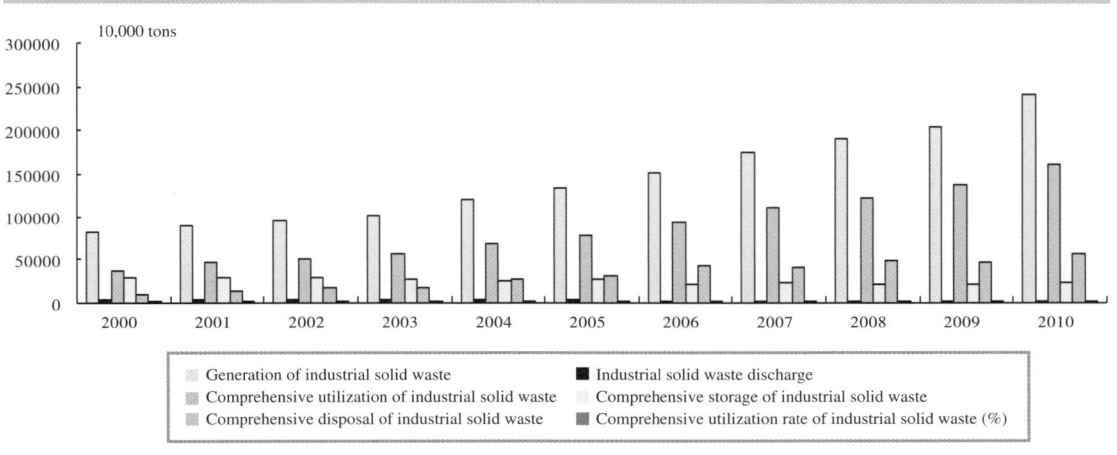

Data: NBS

Urban air quality

In 2009, 612 cities all over China carried out ambient air quality monitoring, and the results showed that the overall urban air quality was good. Among cities at prefecture level and above, the proportion of those whose urban ambient air quality attained the standards was 79.6%. For county-level cities, this proportion was 85.6%. In 2010, although the country's air quality saw a rise, still some cities were confronted with relatively severe environmental pollution.

Results of urban air quality monitoring in 612 cities

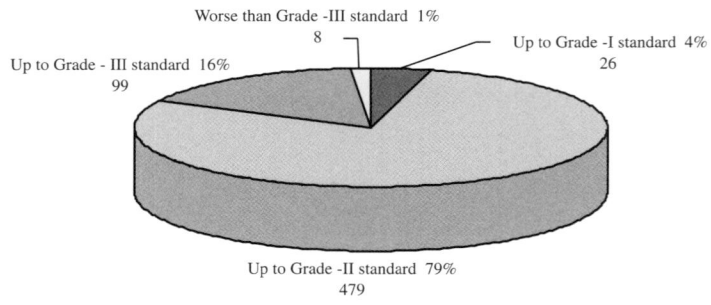

Data: NBS

Among the cities at prefecture level and above, in 2009, only 3.7% of them attained Grade I standard in air quality, 75.9% up to Grade II standard, 18.8% up to Grade III standard, and 1.6% below Grade III standard; in 2010, these four corresponding figures were respectively 3.3%, 78.4%, 16.5% and 1.8%.

In the recent two years, the air quality of 113 key environmental protection cities was somewhat improved. In 2009, 0.9% of the cities attained Grade I standard in air quality, 66.4% up to Grade II standard, and 32.7% up to Grade III standard. Compared with the previous year, the proportion of cities attaining standards rose by 9.8 percentage points. In comparison, in 2010, 0.9% of the cities attained Grade I standard in air quality, 72.6% up to Grade II standard, and 25.6% up to Grade III standard. Compared with last year, the proportion of cities attaining standards increased by 6.2 percentage points.

Proportions of key cities attaining each grade of air-quality standard in 2009

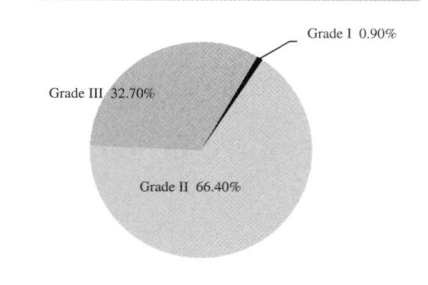

Proportions of key cities attaining each grade of air-quality standard in 2010

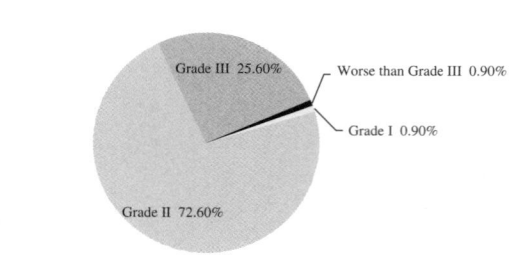

Data: NBS

Industries

Chapter 7   Oil Industry

Section 1: Oil reserves, production and supply in China

Oil strategic reserves

Drawing on the experience of Europe, America and Japan, China's oil reserves system is designed into four hierarchies: state strategic oil reserves, local governmental oil reserves, commercial reserves of state-owned petroleum corporations, and oil reserves of other small and medium-sized companies.

On December 18, 2007, the state oil reserve center was officially founded, aiming to reinforce our construction of strategic oil reserves and improve the system of oil reserve management.

There are three conditions for the selection of an oil reserve base: first, it should be near a deepwater port, railroad line and/or highway network, namely, it should possess excellent transportation and logistics conditions; second, it should be close to large-scale oil refineries so that at certain critical moments the reserve base can produce processed oil to meet the demand; third, it should be near the consumer market.

China is planning to spend 15 years, divided into three phases, to complete the construction of oil reserve bases.

In the first phase, the four strategic oil reserve bases have been put into all-round operation since 2008, which are situated within four coastal regions, i.e. Zhoushan and Zhenhai in Zhejiang Province, Liaoning within Daliang City and Huangdao in Shandong Province. The capacity of crude oil reserves hit 16.40 million m^3 roughly equivalent to 102 million barrels.

In the second phase, China will start to construct eight Phase II strategic oil reserve bases, including Zhanjing and Huizhou in Guangdong Province, Lanzhou in Gansu Province, Jintan in Jiangsu Province, Jinzhou in Liaoning Province and Tianjin Municipality. The total capacity of these Phase II oil strategic reserves amounts to 26.80 million m^3, roughly equal to 168 million barrels. It is predicted to be completed during the years 2012–2013, and by then the total capacity of the state strategic reserves will have been increased by 163%.

The third phase is still being planned. Provinces and cities nationwide are competing with each other for the approval of reserve bases, among which Wanzhou District in Chongqing Municipality, Hainan Province and Caofeidian in Hebei Province are highly likely to be selected as the oil reserve bases for the Phase III project.

Rankings of China's top 11 strategic oil reserve bases in scale

Tianjin Oil Reserve Base (Sinopec Pipeline Company) (Phase II)

This base has been put into construction, and is planned to contain state oil reserve tanks of 5 million m^3 capacity, and commercial oil reserve tanks with a capacity of over 5 million m^3, its total capacity being 10 million m^3. It is predicted to be able to reserve over 6 million tons of oil, and is expected to be the largest-scale oil reserve base in today's China.

Shanshan Oil Reserve Base (EPC Project Division of Xinjiang Oilfield Corporation under PetroChina) (Phase II)

With a total investment of up to 6.5 billion yuan, it will achieve a capacity of 8 million m^3 when completed. At the end of 2008, the Phase-I project, a base with a capacity of 1 million m^3 was formally constructed and put into operation. Meanwhile, crude oil from Kazakhstan was poured into the base as a reserve.

Zhoushan Oil Reserve Base (Sinochem Group) (Phase I, Phase II)

In July 2010, this base passed the combined acceptance by relevant authorities, with a total capacity of 5 million m^3. In 2009, an extra reserve of 2.5 million m^3 was approved to be added as part of the second phase of the oil reserve base construction plan, so that it total capacity will amount to 7.5 million m^3.

Dushanzi Oil Reserve Base (constructed and managed by PetroChina Dushanzi Petrochemical Corporation) (Phase II)

The construction work started in July 2009, marking the overall launch of the Phase II oil reserve base construction. The total planned scale is 5.4 million m^3, with investment totaling 2.65 billion yuan, and is predicted to be completed in July 2011.

Zhenhai Oil Reserve Base (Sinopec) (Phase I)

On December 19, 2007, it passed the state acceptance and became the first among the first batch of oil reserve bases to be handed over for use, and also the largest one in the first phase. Its construction scale amounted to 5.2 million m^3, with a total of 52 oil storage tanks, all of which have by now reserved crude oil.

Huizhou Oil Reserve Base (China National Offshore Oil Corporation (CNOOC)) (Phase II)

Approved by the State Development and Reform Commission (SDRC), Huizhou City in Guangdong Province has made a preliminary plan to build a state oil reserve base with a capacity of 5 million m^3.

Huangdao Oil Reserve Base (managed by Sinopec Pipeline Company, formerly affiliated to State-owned Assets Supervision & Administration Commission) (Phase I)

The main work of the base includes 32 double-deck floating roof tanks with each single having a capacity of 100,000 m³ and the supporting equipments, with a total capacity of 3.2 million m³. It was formally put into operation in November 2008.

Dalian Oil Reserve Base (PetroChina) (Phase I)

It contains 30 oil storage tanks of 100,000 m³, totaling 3 million m³.

Lanzhou Oil Reserve Base (Lanzhou Petrochemical Corporation of PetroChina) (Phase II)

With a total project investment of 2.378 billion yuan, the project mainly includes 30 oil storage tanks with a capacity of 100,000 m³, totaling 3 million m³. All work is expected to have finished and the base put into use by the first half of 2011.

Jinzhou Oil Reserve Base (PetroChina) (Phase II)

PetroChina is planning to build in Jinzhou Development Area a state oil reserve base. The project consists of constructing underground water-sealed oil storage tanks in caverns in Jinzhou, the capacity of which is designed to be 3 million m³, with the planned investment being 2.26 billion yuan.

Jintan Oil Reserve Base (PetroChina) (Phase II)

Jintan City in Jiangsu Province is the emerging "salt city" of China, boasting mineral deposits of rock salt, which possesses the optimal comprehensive index within east China. The exploitation of pure salt has formed a geological condition with a rock salt cavity of 3 million m³. The underground vugs can reserve oil and natural gas.

Oil production and supply

In 2009, China's crude oil output totaled 190 million tons, dropping by 0.4% year on year. Crude oil production witnessed a significant slowdown, a decline for the first time. This was mainly due to the influence of a relatively weakened domestic oil market; however, in 2010, its crude oil output returned to growth with the total crude oil output for the year surpassing the mark of 200 million tons, a year-on-year increase of 7.2%, the highest growth rate within recent years.

Trend of crude oil outputs in China

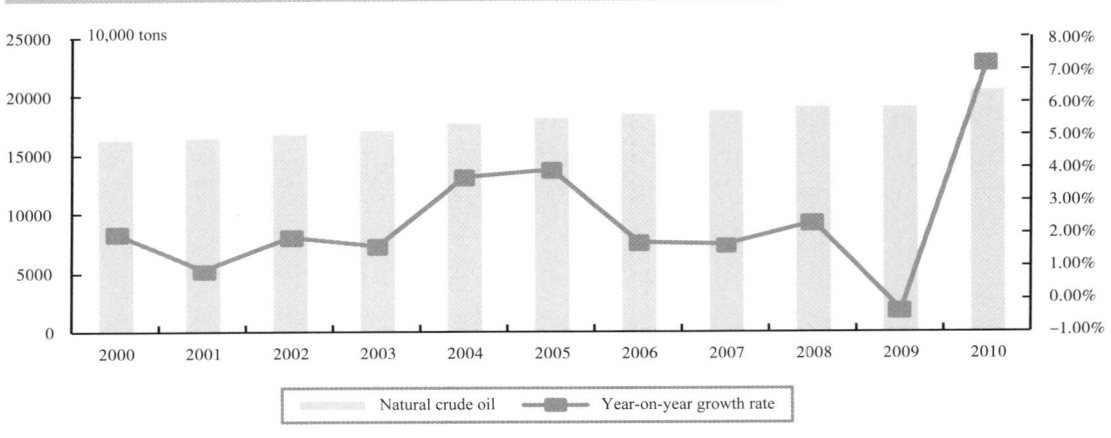

Data source: Statistical Yearbook

In 2009, Heilongjiang possessed China's largest oilfield, Daqing Oilfield, whose production maintained its leading status in China, though witnessing a slight drop.

In 2010, the largest oil-producing province was still Heilongjiang. However, as new reserves were constantly discovered and explored, and traditional oil production zones kept weakening as well, the crude oil outputs of Tianjin and Shaanxi surpassed those of Xinjiang and Shandong, ranking the second and the third largest oil production zones respectively, gradually approximating to the largest.

Rankings of China's top 10 provinces and regions in crude oil output in 2009

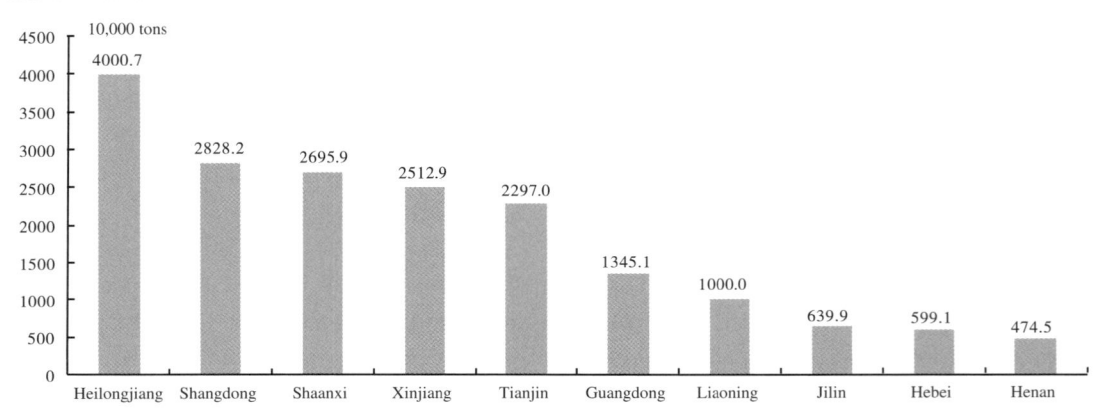

Data source: Statistical Yearbook

Rankings of China's top 10 provinces and regions in crude oil output in 2010

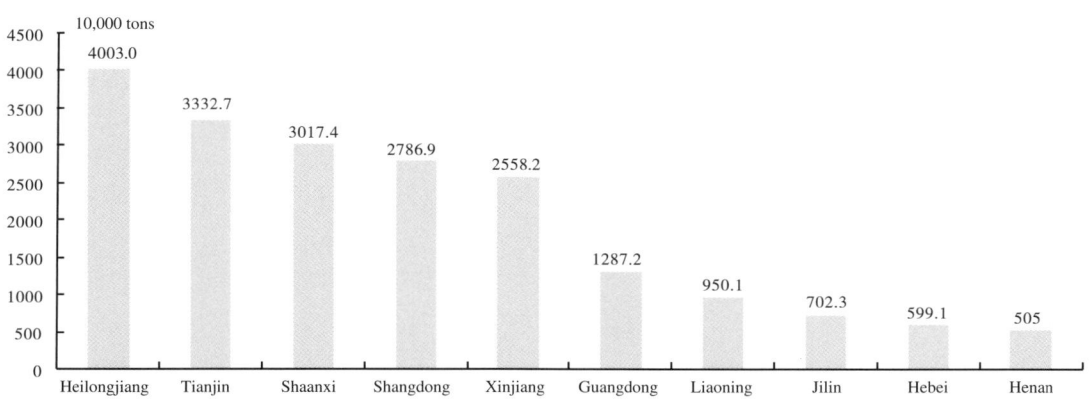

Data source: CEIC

In terms of these 10 largest oilfields, although the crude oil output of Daqing Oilfield has entered a decline period, yet its production still remains stable at the first place. In 2009, its output accounted for 21% of the national total output of crude oil, and 1/3 of the total oil production of China.

Crude oil outputs of China's top 10 oilfields in 2009

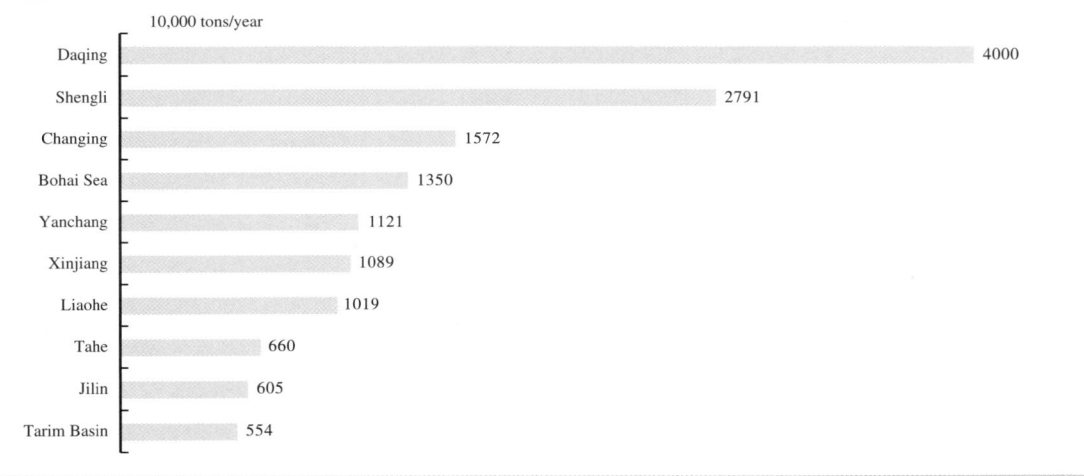

Data source: International Oil Economy

In 2009, the processing capacity of crude oil and processed oil output in China continued to rapidly increase. The crude oil processing capacity for the year hit 370 million tons, a year-on-year increase of 7.9%; the processed oil production was 230 million tons, up 9.4% year on year.

In 2010, influenced by factors such as a relatively rapid growth in oil consumption, an increase in commercial reserves of processed oil, improvement of crude oil processing capacity and a guaranteed profit in processing crude oil, the processing capacity of crude oil broke through for the first time the level of 400 million tons, with a growth rate hitting 13.4%; and the processed oil output was 250 million tons, up 10.9% year on year.

China's processing capacity of crude oil

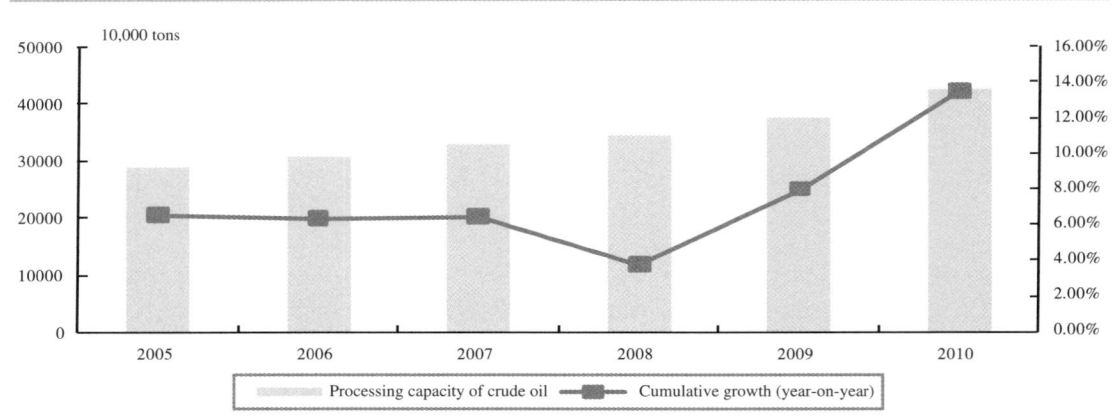

Data source: NBS

Processed oil output

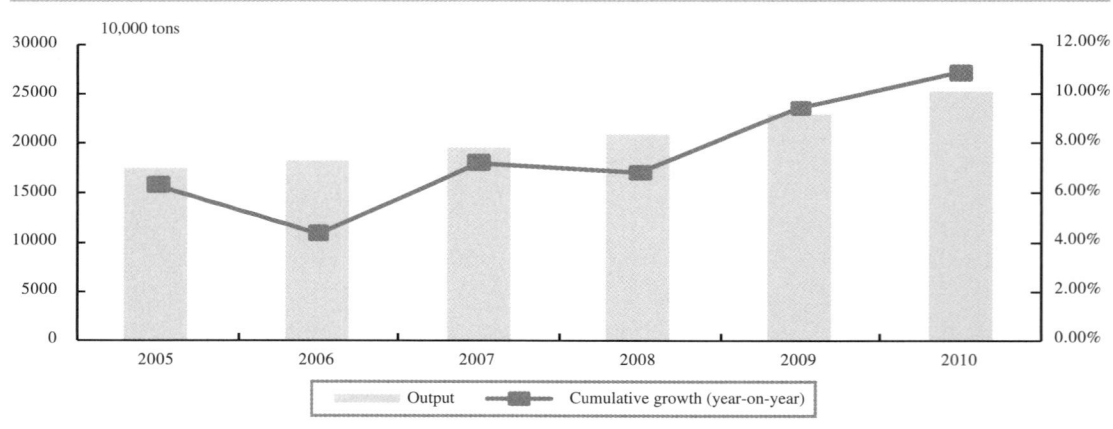

Data source: NBS

Looking at the trend for the whole year, in 2009, crude oil output showed a tendency of declining first and then rising later with a relatively significant amplitude of fluctuation; by contrast, in 2010, the monthly crude oil output surpassed for the first time the mark of 17 million tons in May, constantly hitting a record surge.

Crude oil output of China in 2009

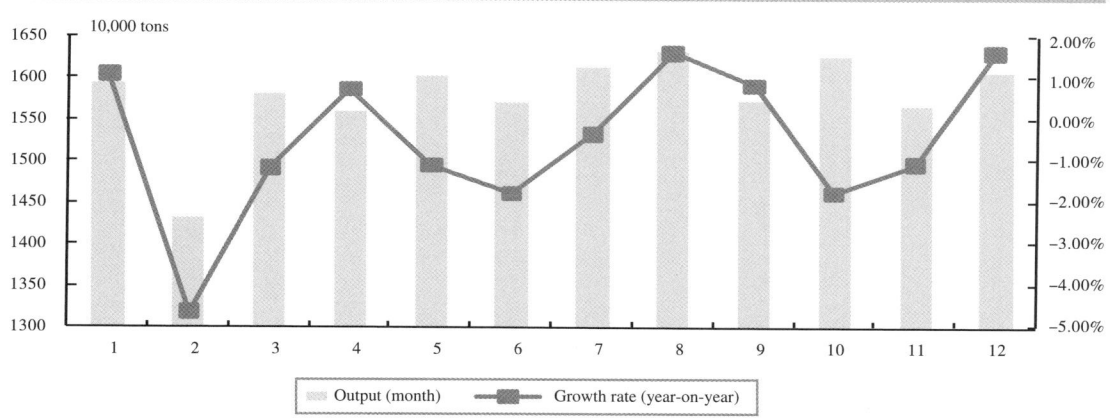

Data source: NBS

Crude oil output of China in 2010

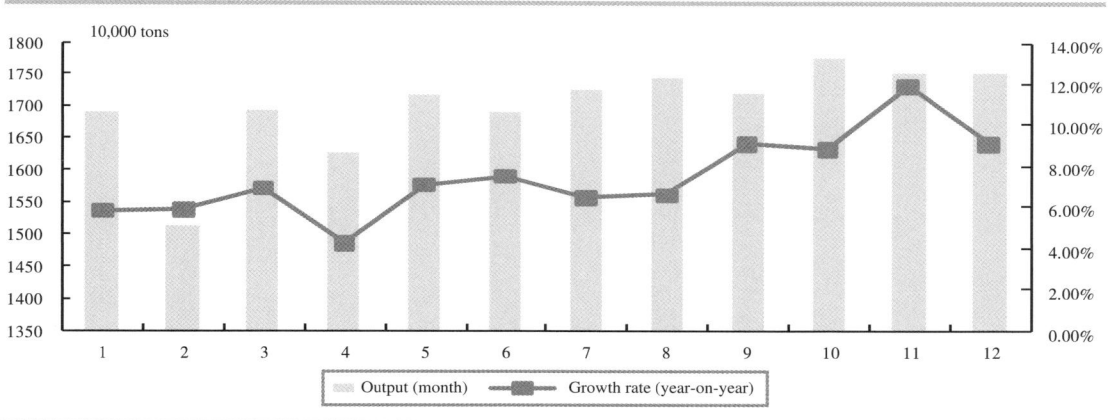

Data source: NBS

In terms of outputs of various oil products, their growth rates are different from each other.

In 2009, the growth rates of gasoline, kerosene and asphalt witnessed a rise, while those of diesel and fuel oil underwent a decline. Within the whole year, the national asphalt output was 23.544 million tons, a year-on-year increase of 54.6%, that of kerosene was 14.794 million tons, rising by 27.0%, and for gasoline, 71.948 million tons and 13.1% respectively, all of which were far higher than the growth rate in recent years. Among them, the diesel output was 141,268 thousand tons, up 6.0% year on year, lower than the growth rate in recent years; the fuel oil output was 16,895 thousand tons, dropping by 19.0% year on year.

In 2010, diesel was severely affected by the financial crisis, and the growth rate of its output was relatively large; in contrast, for such products as gasoline that were slightly affected by the financial crisis, their output growth rate was relatively small.

Outputs of gasoline, kerosene, asphalt, diesel and fuel oil in China (10,000 tons)

Index	Output: gasoline: cumulative value	Output: gasoline: cumulative growth (year-on-year)	Output: kerosene: cumulative value	Output: kerosene: cumulative growth (year-on-year)	Output: oil asphalt: cumulative value	Output: oil asphalt: cumulative growth (year-on-year)	Output: diesel: cumulative value	Output: diesel: cumulative growth (year-on-year)	Output: fuel oil: cumulative value	Output: fuel oil: cumulative growth (year-on-year)
2000	4134.7	12.8%	872.3	21.4%	468.2	8.9%	7079.6	14.4%	—	—
2001	4154.7	0.5%	789.4	−10.1%	516.1	13.6%	7485.7	3.0%	—	—
2002	4320.8	4.0%	826.1	5.1%	722.8	28.6%	7706.1	2.9%	—	—
2003	4790.9	10.9%	855.3	3.6%	861.3	16.3%	8532.8	10.7%	—	—
2004	5278.0	10.2%	970.8	14.1%	902.6	12.5%	10178.7	19.3%	—	—
2005	5405.3	3.2%	988.6	2.6%	936.0	6.0%	11061.6	9.5%	—	—
2006	5591.4	3.7%	960.0	−2.9%	1238.1	19.3%	11653.4	5.5%	2264.7	−6.3%
2007	5994.0	7.2%	1153.3	19.0%	1384.4	8.4%	12370.2	6.2%	2310.0	8.7%
2008	6347.5	5.8%	1165.4	0.5%	1477.6	3.1%	13323.6	8.0%	2228.7	−8.4%
2009	7194.8	13.1%	1479.5	27.0%	2354.4	54.6%	14126.8	6.0%	1856.7	−19.0%
2010	7685.3	5.1%	1714.7	15.3%	2717.9	10.4%	15887.4	12.0%	2115.4v	11.6%

Data source: NBS

Section 2: China's oil demand and oil prices

In 2009, China's apparent consumption of oil hit 410 million tons, only second to that of the USA, ranking second in the world, with a year-on-year rise of 4.0% and a slowdown in its growth rate over the previous year. Among this, the apparent consumption of crude oil made up 390 million tons, up 6.5% year on year, a faster growth rate than that in the previous year.

In 2010, oil supply and demand in China's domestic oil market witnessed a strong growth, along with a rapid growth in apparent consumption of oil, the total apparent consumption year-around hitting a record of 450 million tons, up 12.2% year on year. Since the coming of the new century, this was the only year, except 2004, that witnessed a double-digit growth rate. Among that, apparent consumption of crude oil was 440 million tons, a year-on-year rise of 13.2%.

Apparent consumption of Oil in China

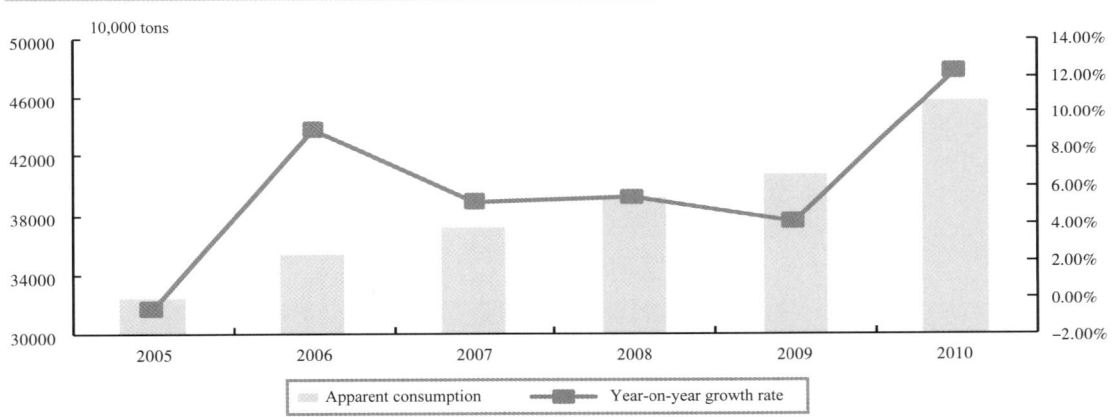

Data source: NBS

Apparent consumption of Crude oil

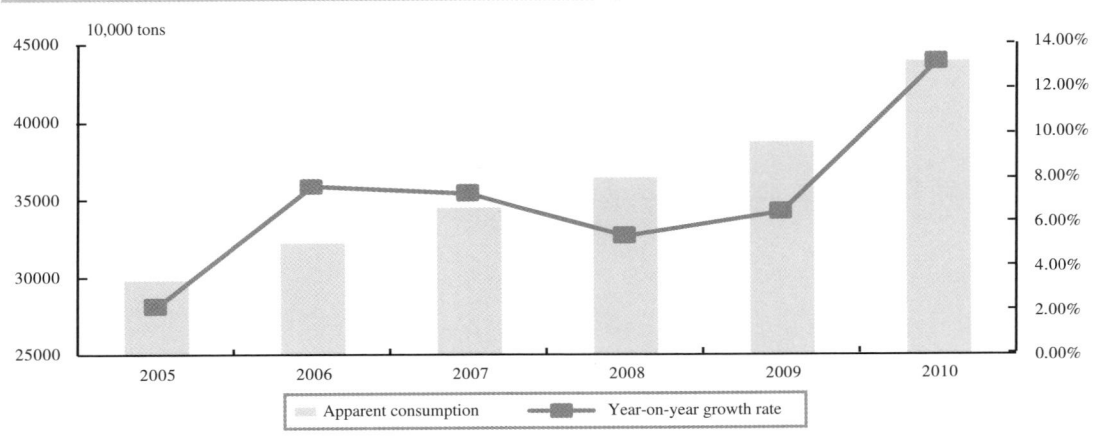

Data source: NBS

Looking at the year-round tendency, in 2009 China's economy declined first and then went up, generally returning to a good status, achieving a GDP growth rate of 9.2%. In close correspondence with the change in the macro economic situation, the year-round oil demand showed a tendency to decline then rise, with an increase in overall demand yet a slowdown in its growth rate. Within the year, in February, apparent consumption of domestic oil was only 27.78 million tons, with its year-on-year decrease rate hitting 9.9%, the lowest level in the recent two years. Later on, state investment and the economic stimulus policy gradually took effect, with apparent consumption of oil constantly rising and hitting a record high.

In 2010, the Chinese economy saw a relatively rapid year-on-year growth, with a trend of slowing down the fluctuation in its growth rate. Correspondingly, China's apparent consumption of oil during the first three quarters witnessed a declining year-on-year growth rate with fluctuation from a high level, then returned to the high level in the fourth quarter.

In 2009, the demands for various oil products experienced quite different changes in growth. For gasoline, kerosene and LPG, their apparent consumption witnessed a year-on-year growth, while for diesel and fuel oil, their apparent consumption saw a year-on-year decline.

In 2010, since oil products faced different influences from the financial crisis, the domestic consumption structure of oil products underwent a change: for kerosene and diesel, the growth rates of apparent consumption were above the double-digit level, while that for gasoline remained relatively small.

Apparent consumption of major oil products in 2010

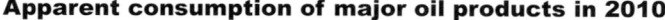

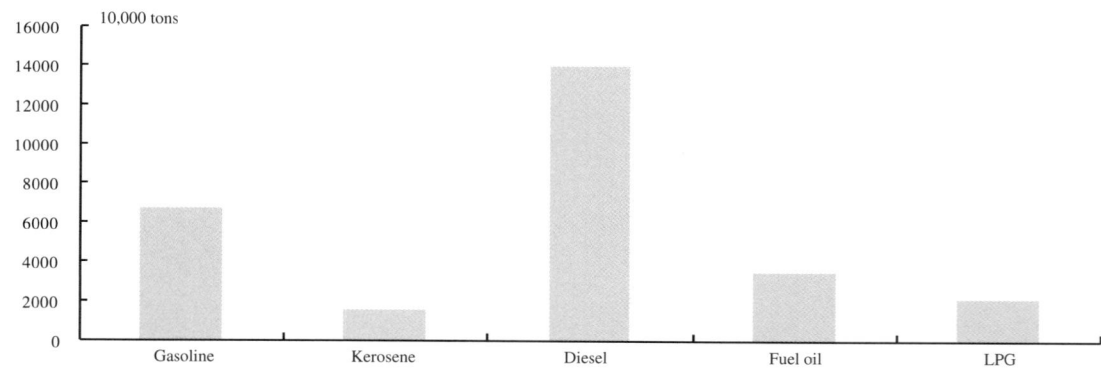

Data source: NBS

Apparent consumption of major oil products in 2010

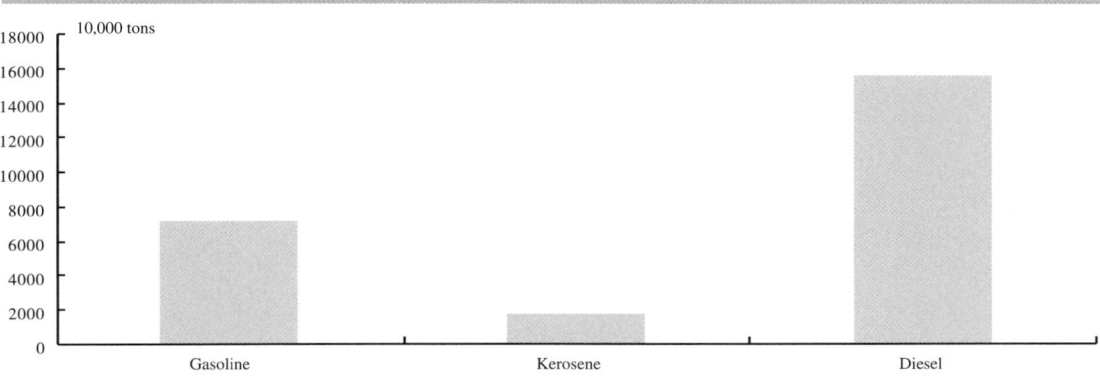

Data source: NBS

Oil prices

In January 2009, China began to put into effect the new pricing mechanism of processed oil: if the margin of change in the average crude oil price in the international market remained above 4% consecutively for 22 work days, it was allowable to adjust the domestic processed oil price.

Relevance between domestic processed oil price and international crude oil price

International crude oil price	Pricing standards of domestic processed oil
Below 80 dollars/barrel	Processed oil price calculated based on normal processing ratio.
Above 80 dollars/barrel	Begin to reduce the processing ratio to a zero processing ratio to calculate the processed oil price.
Above 130 dollars/barrel	Under the principle of considering both the producer and the consumer interests and maintaining a steady operation of the national economy, adopt appropriate tax policies to guarantee the production and supply of processed oil. In principle, the prices of gasoline and diesel both will not be raised or raised a little.

　　In 2009, domestic oil prices witnessed a rise for five times and decline for three times. Generally speaking, the average time interval of these five times of increase was about 48 days, and that of the decline was 28.3 days.

　　Since the middle ten days of February 2010, due to the influence of the global economic recovery, oil demand went up with a constants increase in international market oil price. Our country adjusted the processed oil price for the first time at 0:00 a.m. on April 14 and within the year oil prices were raised three times and lowered once.

Price adjustments of gasoline #93 in Beijing in 2009 and 2010

Date of adjustment	Retail price after adjustment (yuan/liter)	Range of price changes (yuan/liter)
2010.12.22	7.14	0.22
2010.10.26	6.92	0.18
2010.06.01	6.74	−0.18
2010.04.14	6.92	0.26
2009.11.10	6.66	0.38
2009.09.29	6.28	−0.15
2009.09.01	6.43	0.22
2009.07.29	6.21	−0.16
2009.06.30	6.37	0.48
2009.06.01	5.89	0.33
2009.03.25	5.56	0.23
2009.01.15	5.33	-0.11

Looking at the adjustment margin, this margin for the Chinese domestic processed oil price remained relatively steady, with each adjustment below the amplitude of fluctuation of the international oil price.

In 2009, China's gasoline price and diesel price cumulatively rose by 1520 yuan and 1390 yuan respectively, both with a growth rate of around 27%, while the crude oil price in the international market in the same period rose from 40 dollars/barrel at the beginning of the year to the present 73 dollars/barrel, with a growth margin hitting 82.5%.

Oil price adjustments in 2009 and 2010

2009 01.15	Gasoline price declined by 140 yuan/ton; diesel price declined by 160 yuan/ton
2009.03.25	Gasoline price rose by 290 yuan/ton; diesel price rose by 180 yuan/ton
2009.06.01	Prices of gasoline and diesel both rose by 400 yuan/ton
2009.06.30	Processed oil price rose by 600 yuan/ton
2009.07.29	Processed oil price declined by 200 yuan/ton
2009.09.02	Processed oil price declined by 300 yuan/ton
2009.09.30	Prices of gasoline and diesel declined both by 190 yuan/ton
2009.11.09	Prices of gasoline and diesel both rose by 480 yuan/ton
2010.04.14	Prices of gasoline and diesel both rose by 320 yuan/ton
2010.05.31	Prices of gasoline and diesel declined by 230 yuan/ton and 220 yuan/ton respectively
2010.10.25	Prices of gasoline and diesel rose by 230 yuan/ton and 220 yuan/ton respectively
2010.12.21	Prices of gasoline and diesel rose by 310 yuan/ton and 300 yuan/ton respectively

Gasoline and diesel prices in 2009

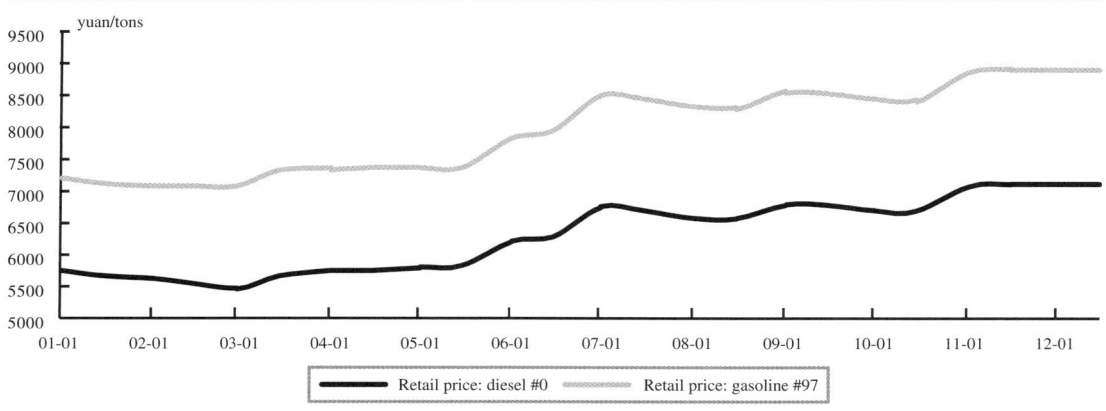

Data source: Ministry of Commerce

Gasoline and diesel prices in 2010

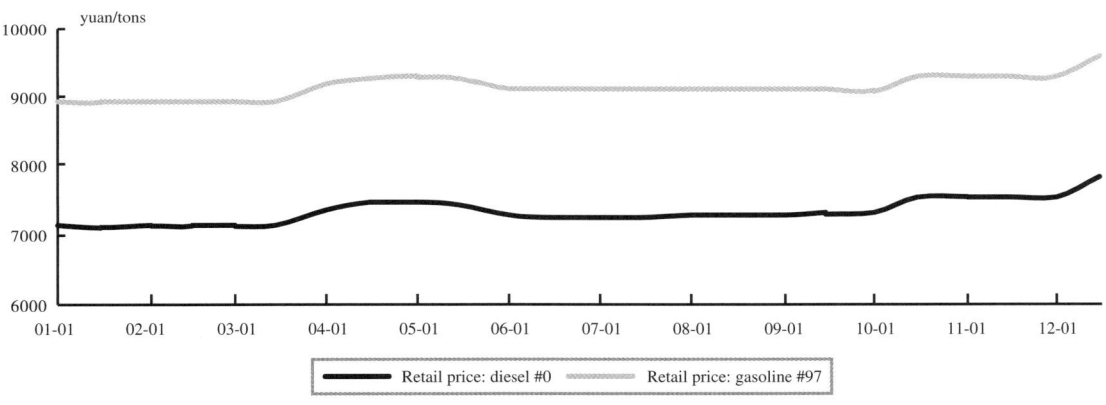

Data source: Ministry of Commerce

Section 3: Analysis of oil imports/exports in China

In 2009, domestic crude oil production declined and oil demand continued to grow, which led to the reality of 256.424 million tons of year-round oil imports, with a year-on-year rise of 11.4%; 39.166 million tons of exports with a year-on-year rise of 2.9%; 217.258 million tons of net imports, a year-on-year rise of 8.3%, with its growth rate falling by 3 percentage points over the year 2008.

In 2010, although domestic crude oil output witnessed a substantial growth, it still fell behind that of oil consumption, causing a significant increase in oil imports, with the total net imports for the year being 255.255 million tons, a substantial growth of 6.6% year on year.

Oil imports/exports in China

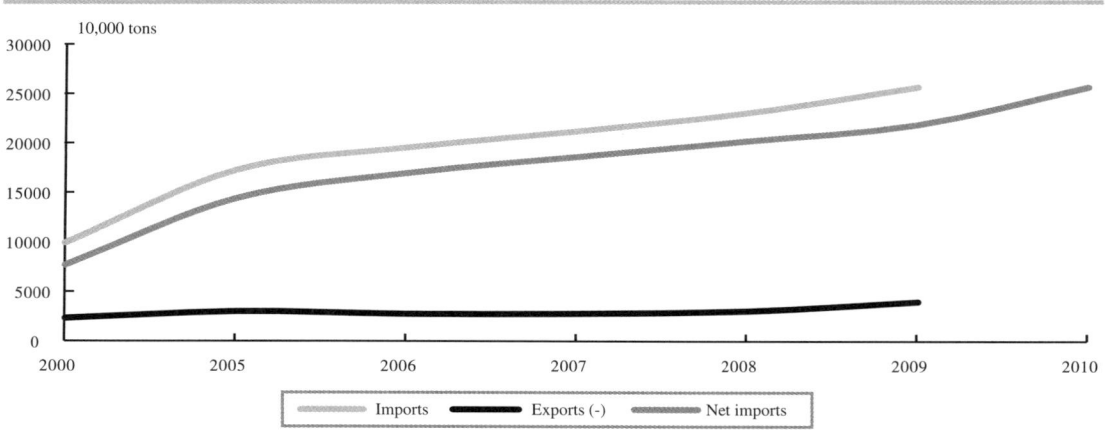

Data source: Energy Statistical Yearbook

In 2009, Chinese domestic crude oil imports witnessed the eighth consecutive year of growth, with its year-round imports surpassing the mark of 200 million tons for the first time, that is, 240 million tons, up 13.9% over the previous year, and making up 80.9% of total oil imports. The fact that a large-scale newly added refining capacity had been put into operation led to a higher growth rate of processed oil supply than that of its demand. Meanwhile, a drop in imports and an increase in exports made processed oil net imports decline by 45%, down to 11.92 million tons, the lowest level within the recent seven years.

In 2010, Chinese domestic net imports of crude oil were 236 million tons, a year-on-year rise of 18.9%, its growth rate being 6 percentage points higher than that of the previous year, making up 92.6% of total oil net imports. In contrast, for processed oil, under the

China's imports/exports of crude oil

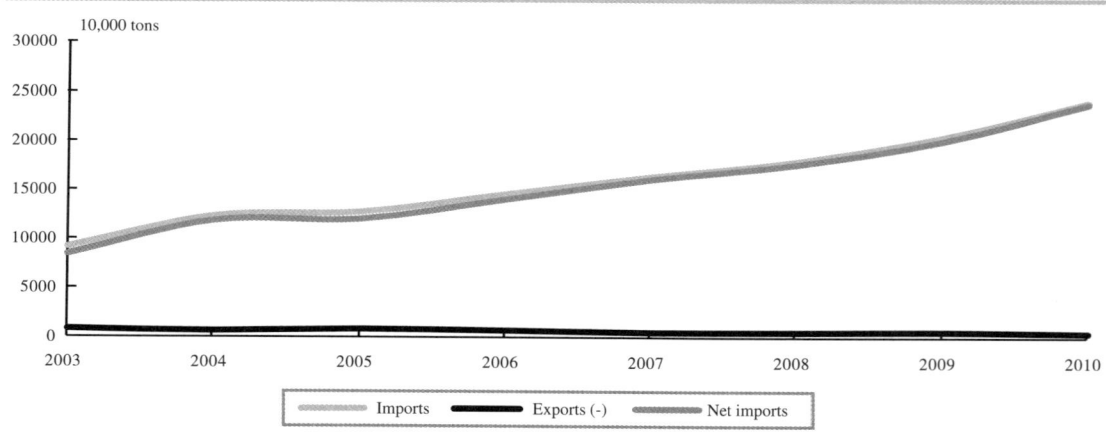

Data source: General Administration of Customs (China)

China's imports/exports of processed oil

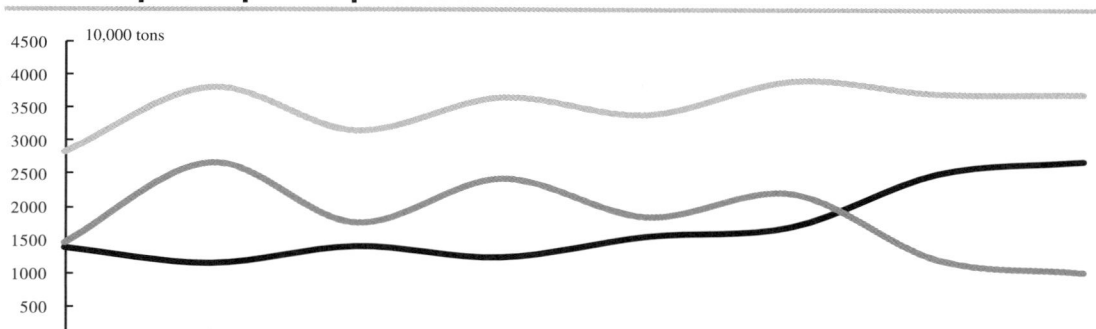

Data source: General Administration of Customs (China)

influence of an unusual fluctuation in domestic market demand, its exports witnessed a significant growth, with a growth rate higher than that of its imports by 7.6 percentage points. Its total net imports of processed oil for the year were 10 million tons, down 16.1% year on year.

Looking at the year-round trend, in 2009, influenced by the trend of domestic oil demand, oil imports saw a tendency of descending first and then going up. In February, oil net imports were only 13.47 million tons, reaching the lowest level since February 2007, with its declining margin hitting 15.5%. Starting from March, with a recovered oil demand, oil imports rebounded substantially; while in 2010, oil net imports saw a trend of fluctuating growth, hitting new highs in March, April, June and September.

In terms of different oil products, diesel net exports in 2009 hit a record high of 2.670 million tons; gasoline net exports also roared from less than 50,000 tons in 2008 to 4.9 million tons; imports of aviation kerosene exceeded exports for the first time since 1992; the newly added ethylene capacity put into operation gave rise to a substantial increase in the demand for petrochemical materials. Consequently, China became for the first time a net importer of naphtha within the recent 10 years, with a total of 1.796 million tons. Due to an increase in residential demand, liquefied petroleum gas (LPG) imports rebounded to 4.091 million tons, equivalent to that in 2007; fuel oil was still the primary oil product in net imports for China, its year-round net imports hitting 15.379 million tons, up 7.1% year on year.

In 2010, the sum total imports of gasoline, kerosene and diesel for the year were 8.3 million tons, up 3.7% year on year; their total exports were 15.397 million tons, rising by 3.5% year on year. Among that, there were no gasoline imports; exports of kerosene exceeded imports at one point; during the months January to October net exports of kerosene hit 295,000 tons. As for other oil products, in 2010 imports of naphtha and petroleum asphalt soared. The year-round net imports of petroleum asphalt amounted to 5.408 million tons, a significant rise of 48.4% year on year. Following crude oil and fuel oil, it remained stable at third place; LPG net imports were 2.279 million tons, significantly decreasing by 29.7% year on year, yet remaining fourth among oil products; lubricant (including lubricant base oil) rose to become the fifth largest net imported oil product.

Statistics of China's oil imports and exports in 2009

Product name	Imports			Exports			Net imports	
	2008	2009	Growth rate	2008	2009	Growth rate	2008	2009
Crude oil	17889.30	20378.89	13.9	373.34	518.40	38.9	17515.96	19860.50
Processed oil	3886.74	3696.39	−4.9	1702.93	2504.55	47.1	2183.81	1191.84
Gasoline	198.70	4.44	−97.8	203.55	494.31	142.8	−4.85	−489.88
Naphtha	77.27	265.12	243.1	151.37	85.47	−43.5	−74.10	179.64
Aviation kerosene	624.91	576.22	−7.8	533.05	594.49	11.5	91.86	−18.27
Light diesel	624.75	183.68	−70.6	62.86	450.71	617.0	561.89	−267.02
Fuel oil	2160.09	2400.35	11.1	724.58	862.45	19.0	1435.51	1537.90
Fuel oils #0–7	1788.76	2090.50	16.9	478.33	647.56	35.4	1310.43	1442.95
Other fuel oils	371.33	309.85	−16.6	246.25	214.89	−12.7	125.08	94.96
Lubricant	24.42	23.65	−3.1	11.60	10.06	−13.3	12.82	13.60
Lubricant base oil	136.93	188.48	37.7	12.75	5.41	−57.6	124.18	183.08
Other processed oils	39.67	54.44	37.2	3.17	1.65	−48.0	36.50	52.79
Liquefied petroleum gas (LPG)	260.02	409.14	57.3	67.52	84.89	25.7	192.50	324.25
Other oil products	415.74	603.96	59.7	241.10	202.39	−16.1	174.64	461.58
Paraffin wax	0.83	0.85	2.6	57.72	54.23	−9.2	−58.89	−53.38
Petroleum coke	92.13	329.67	257.8	179.79	139.99	−22.1	−87.66	189.68
Petroleum asphalt	322.78	333.44	3.3	1.59	8.17	414.0	321.19	325.27
Total	22451.81	25148.39	12.0	2384.89	3310.22	38.8	20066.91	21838.17

Data source: the General Administration of Customs

In 2009, the FTDR both of oil and crude oil surpassed the 50% mark. The year-round total net imports of oil were 217.258 million tons; apparent consumption of oil hit 406.748 million tons, with the FTDR rising from 51.3% in 2008 to 53.4%. As for crude oil, the net imports were 198.72 million tons; its apparent consumption hit 388.209 million tons, with an FTDR of 51.2%.

In 2010, the FTDR of oil and crude oil in China registered a new record since surpassing the 50% warning line for the first time one year ago. The year-round apparent consumption of oil totaled 449 million tons, with the oil FTDR above 55%. For crude oil, its net imports were 236.28 million tons, and the apparent consumption totaled 439.296 million tons, with an FTDR of 53.8%.

China's oil foreign trade dependency ratio

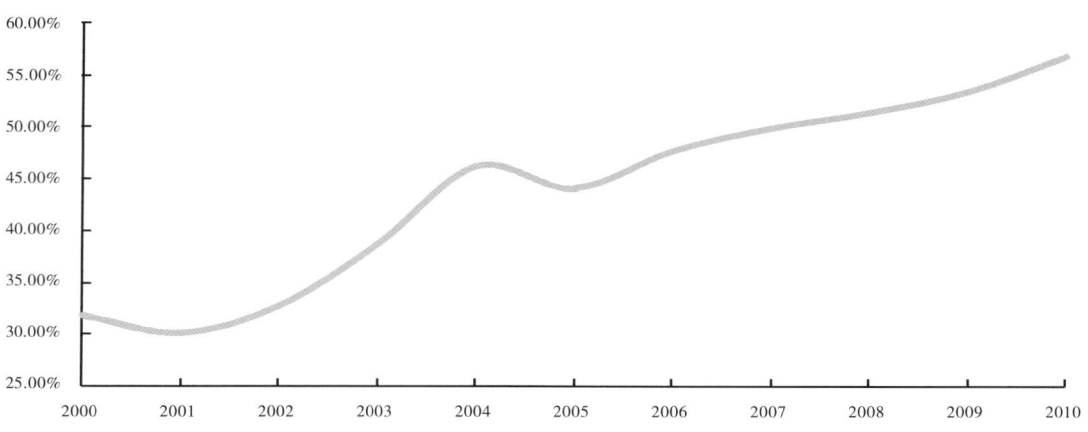

Data source: General Administration of Customs (China)

China's crude oil foreign trade dependency ratio

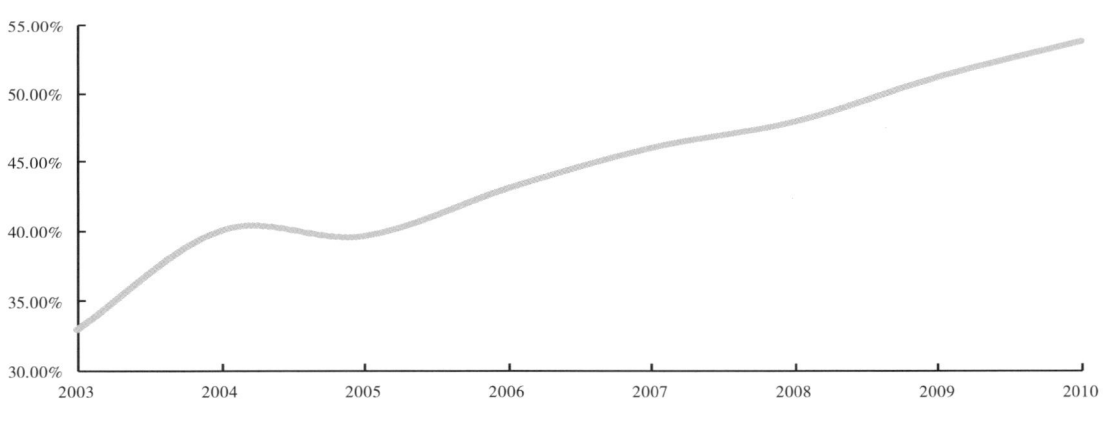

Data source: General Administration of Customs (China)

Section 4: Analysis of China's petrochemical industry

In 2009, there were in total 34,600 enterprises above the scale characteristic of the national oil and chemical industry, their total output value hitting 6.63 trillion yuan, up 0.3% year on year. It accounted for 12.1% of the total national industrial output value, and increased by 10.1% year on year in industrial added value. In comparison, their total sales output value was 6.35 trillion yuan, rising by 0.15%. Over 20 kinds of staple products remained in the front ranks the world over in terms of output, among which major petrochemical products led the world in consumption.

In 2010, pulled by the increase in demand and price, the economic growth of the petrochemical industry continued to speed up, with its output value hitting a record high.

During the months January to November, the enterprises above the whole industrial scale were 36,800, and the total output value of the industry was 7.99 trillion yuan, up 34.6% year on year, among which the output value of new products was 461.66 billion yuan, up 46.2% year on year.

In 2009, the total output value of the chemical industry was 3.93 trillion yuan, up 9.7% year on year, achieving a relatively rapid growth; the total oil output value of the natural gas exploitation industry was 749.03 billion yuan, rising by 26.6%; the total output value of the oil refining industry was 1.78 trillion yuan, falling by 4.2%. Looking at the growth margin, in 2009 the added value of the chemical industry rose by 15.9% year on year, that of oil and natural gas by 4.8%, and that of the oil refining industry by 5.2%, with the chemical industry becoming the major driving force of economic growth.

Statistics on economic operation of China's petrochemical industry in 2009

Industry	Total output value (trillion yuan)	Year-on-year growth (%)	Industrial added value year on year (%)
Chemical industry	3.96	9.7%	15.9%
Oil & natural gas exploitation industry	0.75	26.6%	4.8%
Oil refining industry	1.78	4.2%	5.2%

In 2009, the industrial economy witnessed a big V-shaped trend: influenced by the financial crisis, it reached a trough at the very beginning of 2009. Driven by the national micro policy, it gradually rebounded and recovered from March, and in December it reached a record in output value. From January 2008 to December 2009, the growth margin of output value of the petrochemical industry showed a V shape, which indicated that the petrochemical industry had emerged from the bottom and progressed forward for the better.

The growth of industry investment tended to slow down. In 2009, the whole-industry fixed asset investment was 1.01 trillion yuan, up 12.9% year on year and down 14.2 percentage points over the previous year, which was lower than the average growth rate (30.1%) of national fixed asset investment by 17.2 percentage points in the same period.

In 2010, the whole industry achieved 1.15 trillion yuan in fixed assets investment, with a growth rate of 13.8%, rising by 0.9 percentage points over the previous year, yet lower than the growth rate (23.8%) of city-and-town fixed asset investment by 10 percentage points in the same period.

China's petrochemical industry investment from 2006 to 2010

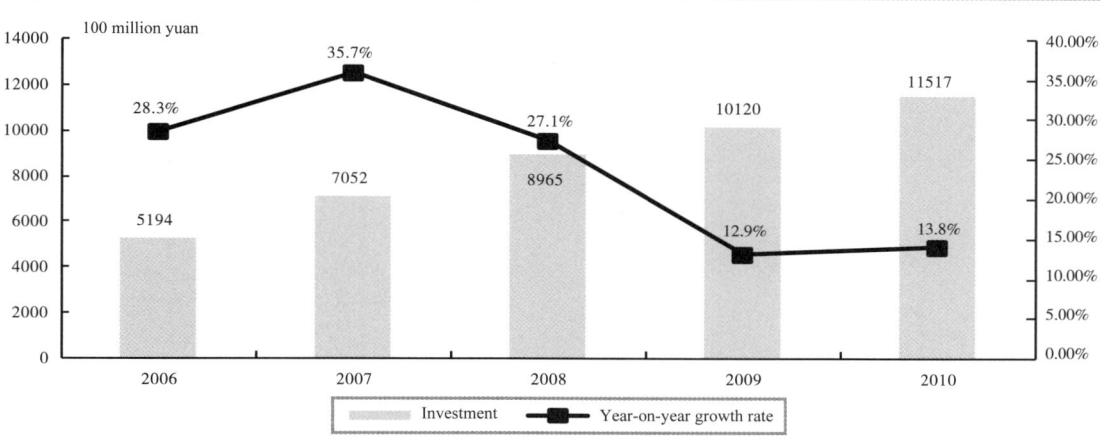

Data source: China Petroleum and Chemical Industry Federation

Total imports/exports of China's petroleum and chemical industry during the years 2006–2010

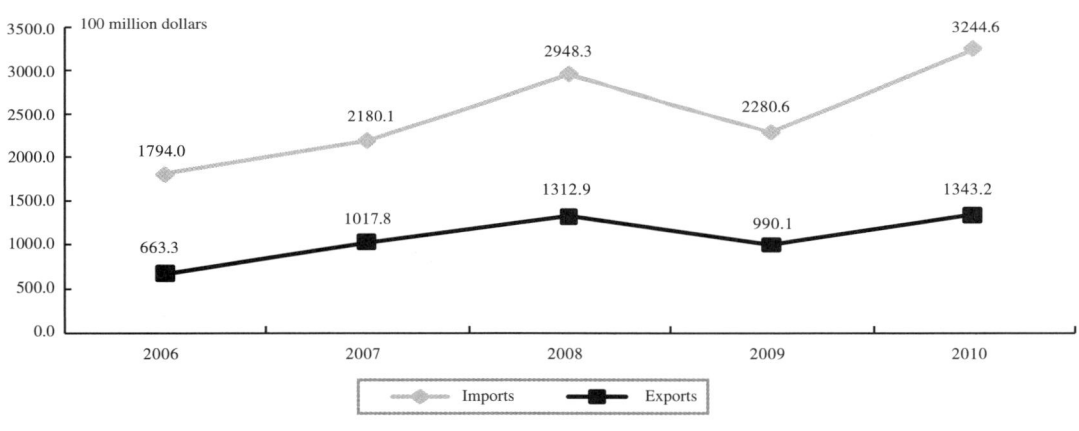

Data source: NBS

Chapter 8   Natural Gas Industry

Section 1: Reserves, production and supply of natural gas in China

In recent years, the natural gas output in China has continued to maintain a high-level growth. In 2009, the total production of natural gas hit 82.99 billion m^3 rising by 6.91 billion m^3 over the previous year, with a year-on-year growth rate of 7.7%, slowdown slightly. In 2010, this total production changed into 96.76 billion m^3, substantially rising by 13.5% year on year, with its growth rate higher than that of last year by 5.8 percentage points.

China's natural gas output

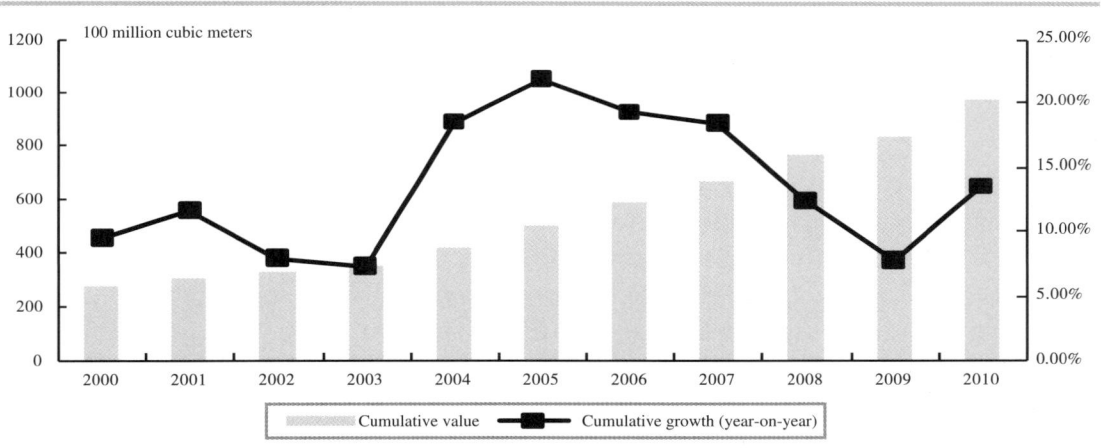

Data source: Statistical Yearbook

Looking at different provinces, among 21 provincial-level administrative regions whose national natural gas achieved large-scale production, in 2009 ranking first was Xinjiang, the big region of traditional resources, followed by Sichuan Province and Shaanxi Province respectively. The natural gas outputs of these three provinces and regions were quite similar, remaining stable in the front ranks of the country, with their sum total making up 75.7% of our national total. In 2010, the total natural gas output of these three regions continued to increase, accounting for 74.9% of the national total.

China's top 10 provinces and regions in natural gas output in 2009 (unit: 100 million cubic meters)

	Output in 2009	Year-on-year growth rate	Share of the national total
Xinjiang	245.4	4.0%	29.6%
Sichuan	193.6	−0.3%	23.3%
Shaanxi	189.5	31.4%	22.8%
Guangdong	58.4	−3.9%	7.0%
Qinghai	43.1	−1.3%	5.2%
Heilongjiang	30.0	10.8%	3.6%
Henan	29.6	163.6%	3.6%
Tianjin	14.3	2.1%	1.7%
Jilin	11.6	33.0%	1.4%
Hebei	10.9	24.4%	1.3%

Data source: Energy Statistical Yearbook

Among all the oil and gas areas, three large gas areas, Sichuan-Chongqing, Tarim Basin and Changqing, were still the major gas fields, with their yearly natural gas outputs each approximating to 20 billion m³. Among these three, Changqing gas field boasts a relatively higher growth rate in natural gas output.

Section 2: Demand for and price of natural gas in China

In recent years, China's natural gas market has welcomed a large-scale development, with the natural gas consumption growing at a double-digit rate.

In 2009, China's natural gas consumption hit 89.52 billion m³ up 10.1% year on year, with a growth rate same as that in the previous year. Among primary energy consumption, that of natural gas still boasted the highest growth rate, the main reasons being the increase in the output of domestically produced natural gas and the scale of LNG imports.

In 2010, China still saw a rapid growth in natural gas consumption, its output totaling 106 billion m³, up 19.5% year on year.

Natural gas consumption in China

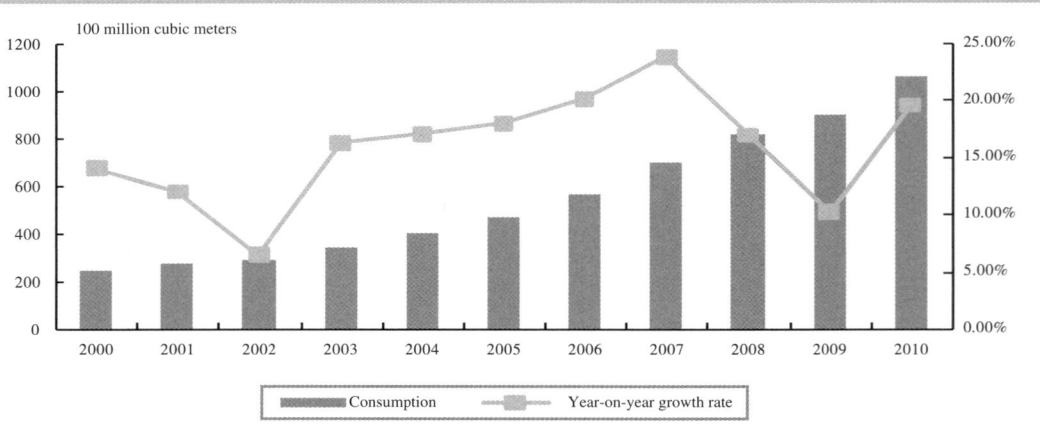

Data source: Statistical Yearbook

During recent years, compared with natural gas production, the growth rate of demand surpassed it, thus leading to a natural gas shortage.

Comparison of growth rates of natural gas production and consumption in China

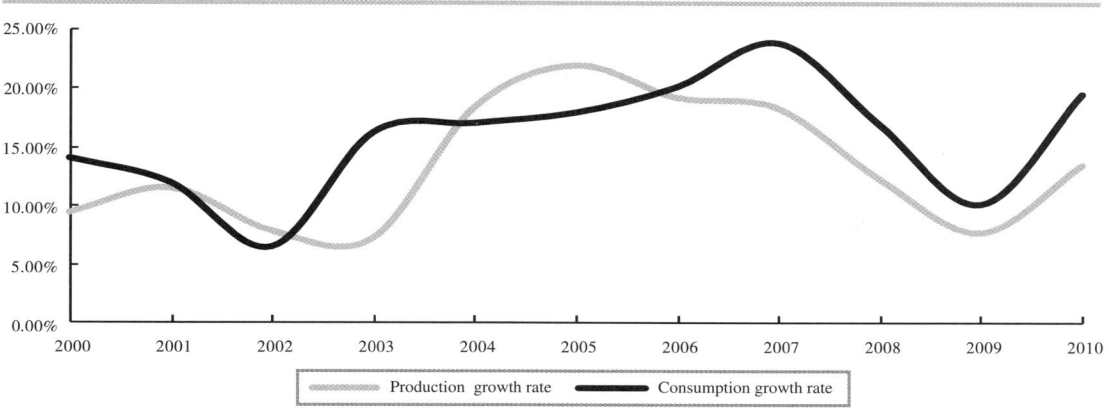

Data source: Statistical Yearbook

Looking at the regional distribution, natural gas consumption was still concentrated in four large areas: the Sichuan-Chongqing area, the area around the Bohai Sea, the Changjiang River Delta and the southeastern coastal area. In 2009, there were 24 provinces/municipalities and regions with their gas consumption surpassing 1 billion m³, and 15 provinces/municipalities with their gas consumption over 2 billion m³. Among them, Sichuan still remained first in terms of yearly natural gas consumption, hitting 12.7 billion m³ and making up above 10% of the national total natural gas consumption.

Natural gas consumption in various provinces and regions of China in 2009
(unit: 100 million cubic meters).

Region	Consumption in 2009	Share of national total
Beijing	69.4	7.2%
Tianjin	18.1	1.9%
Hebei	23.1	2.4%
Shanxi	13.8	1.4%
Inner Mongolia	44.3	4.6%
Liaoning	16.4	1.7%
Jilin	16.7	1.7%
Heilongjiang	30.0	3.1%
Shanghai	33.5	3.5%
Jiangsu	63.4	6.5%
Zhejiang	19.3	2.0%

(Continued)

Region	Consumption in 2009	Share of national total
	(Continued)	
Anhui	9.8	1.0%
Fujian	8.5	0.9%
Jiangxi	2.6	0.3%
Shandong	40.2	4.2%
Henan	41.5	4.3%
Hubei	16.5	1.7%
Hunan	10.2	1.1%
Guangdong	112.9	11.7%
Guangxi	1.2	0.1%
Hainan	24.9	2.6%
Chongqing	49.5	5.1%
Sichuan	127.0	13.1%
Guizhou	4.2	0.4%
Yunnan	4.5	0.5%
Shaanxi	50.0	5.2%
Gansu	12.4	1.3%
Qinghai	24.6	2.5%
Ningxia	12.0	1.2%
Xinjiang	67.9	7.0%

Data source: Energy Statistical Yearbook

From the perspective of the consumption structure, the consumption structure of natural gas continues to be optimized. Along with the constant increase in city gas demand is the fall in the demand for industrial fuels and gas used for chemical engineering. In 2009, among the city gas consumption, that of natural gas totaled 21.3 billion m^3, with a 24.0% share of the total consumption; the consumption for industrial fuel use was 34.3 billion m^3, making up 38.7% of the total; that for chemical engineering hit 18 billion m^3, making up 20.3% of the total; and that for electricity generation was 15 billion m^3, accounting for 16.9% of the total consumption.

Natural gas consumption structure of China in 2009

	Consumption (100 million cubic meters)	Share
City gas	213	24.0%
Industrial fuel gas	343	38.7%
Gas for chemical engineering	180	20.3%
Gas for electricity generation	150	16.9%

Data source: Statistical Yearbook

China's natural gas price has reached its trough. Compared with alternative energy, its price was lower than not only that of gasoline, but even that of coal. In comparison with natural gas prices in other countries, there exists a price inversion between the prices abroad and at home. The relatively lower price cannot fully reflect the value of natural gas, and thus fails to play its role in reflecting the degree of scarcity in energy and encouraging energy-saving utilization.

The comparison between different kinds of energy in terms of the same heating value also indicates a lower natural gas price. The data shows that under the condition of a same heating value, our country's natural gas price is only equivalent to 25–30% that of crude oil, 30–40% that of electricity, while the international natural gas price is equivalent to 60–80% that of crude oil.

In China, the natural gas price differs in different provinces and regions even though used in the same fields. Generally speaking, the price level of natural gas shows a pattern of high in the east while low in the west.

With the gradual recovery of oil prices, there appears an increasingly widening gap between the ex-factory price of natural gas and that of alternative energy like crude oil. In particular, it's universally expected that as the project of the pipeline natural gas importing from Central Asia was put into operation, our country would issue a reform scheme on the natural gas price in 2009. However, for a variety of reasons, the price adjustment scheme was modified again and again and was not finally issued.

In 2010, the natural gas price finally made its first step. Starting from 0:00 a.m. on June 1, the base (ex-factory) price of domestically produced overland natural gas was raised by 230 yuan per 1000 m^3; enlarging the margin of price fluctuation, it uniformly elevated the margin of allowable fluctuation for the base (ex-factory) price up to 10% with no limit in the downward fluctuation. It meant that both suppliers and consumers can negotiate to decide the specific natural gas price under the condition that it is no more than 10% of the base (ex-factory) price.

Section 3: Typical oil and natural gas enterprises in China

In 1949, crude oil output in China was only 120,000 tons. In 1955, the Chinese government officially founded the Ministry of Petroleum Industry, fully responsible for the exploitation and development of oil and natural gas resources. Its predecessor was the Petroleum Administration Bureau of Ministry of Fuel Industry that was founded in 1950.

In 1980, the State Council of China Ministry of Petroleum Industry, and then successively set up China National Offshore Oil Corporation (CNOOC), China Petrochemical Corporation and China National Petroleum and Natural Gas Corporation (CNPNG).

In 1998, the State Council of China further restructured the businesses of CNPNG and Sinopec Group and then founded China National Petroleum Corporation (CNPC) and China Sinopec Group.

Presently, China's petrochemical industry shows a pattern of three giants competing against each other balanced in force.

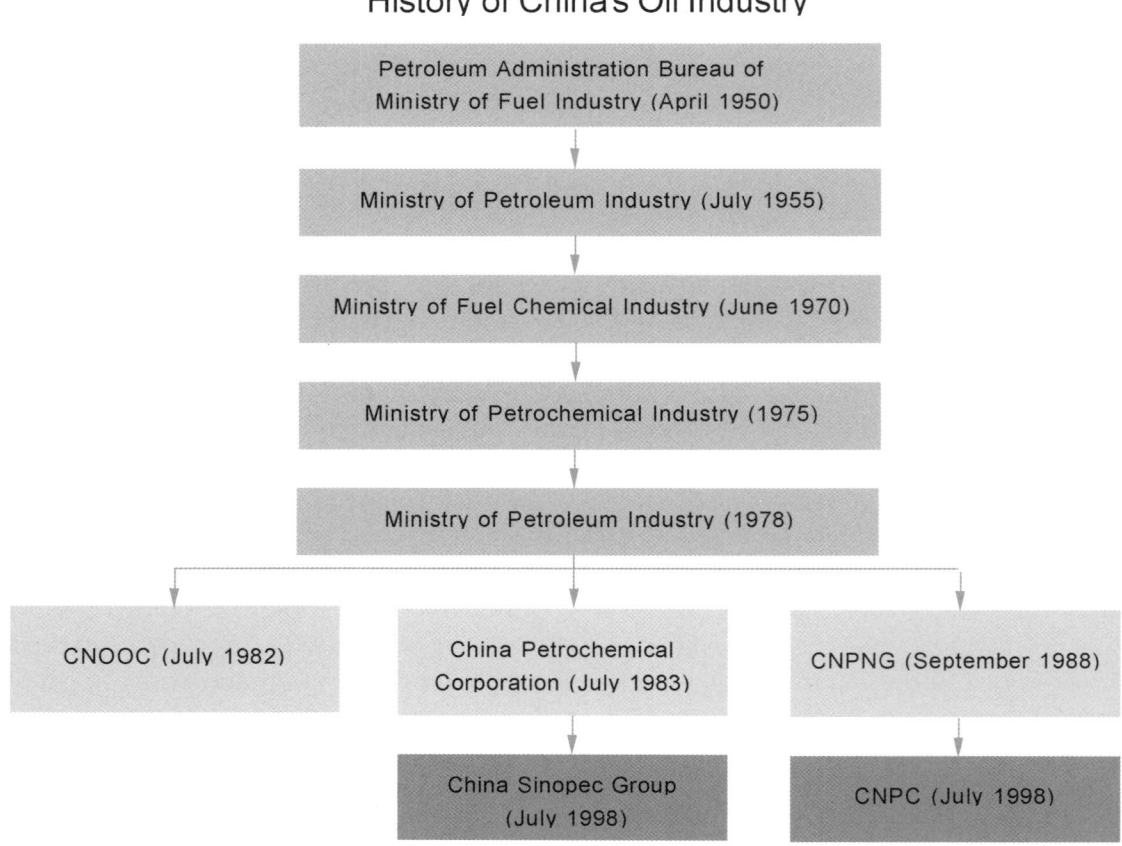

History of China's Oil Industry

China national petroleum corporation (CNPC)

Company overview

China National Petroleum Corporation (CNPC), founded on the basis of the former China National Petroleum and Natural Gas Corporation, is a mega group of petroleum and petrochemical enterprises. Its major operations include oil and gas businesses engineering services, petroleum engineering construction, petroleum equipment manufacturing, financial services and new energy development.

Equity structure

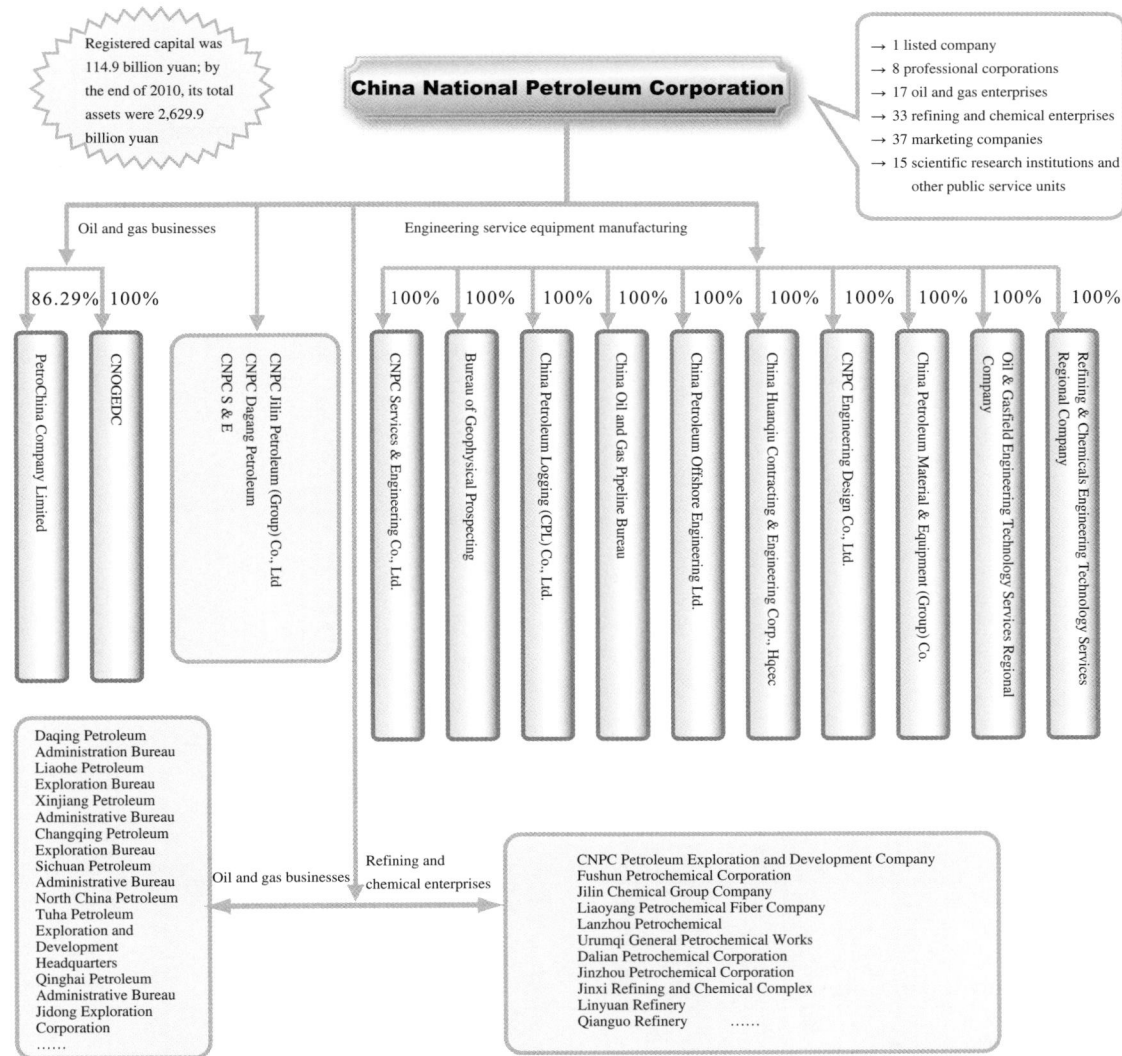

Registered capital was 114.9 billion yuan; by the end of 2010, its total assets were 2,629.9 billion yuan

China National Petroleum Corporation

→ 1 listed company
→ 8 professional corporations
→ 17 oil and gas enterprises
→ 33 refining and chemical enterprises
→ 37 marketing companies
→ 15 scientific research institutions and other public service units

Oil and gas businesses

Engineering service equipment manufacturing

86.29%   100%   100%   100%   100%   100%   100%   100%   100%   100%   100%   100%   100%

PetroChina Company Limited

CNOGEDC

CNPC Jilin Petroleum (Group) Co., Ltd
CNPC Dagang Petroleum
CNPC S & E

CNPC Services & Engineering Co., Ltd.

Bureau of Geophysical Prospecting

China Petroleum Logging (CPL) Co., Ltd.

China Oil and Gas Pipeline Bureau

China Petroleum Offshore Engineering Ltd.

China Huanqiu Contracting & Engineering Corp., Hqcec

CNPC Engineering Design Co., Ltd.

China Petroleum Material & Equipment (Group) Co.

Oil & Gasfield Engineering Technology Services Regional Company

Refining & Chemicals Engineering Technology Services Regional Company

Daqing Petroleum Administration Bureau
Liaohe Petroleum Exploration Bureau
Xinjiang Petroleum Administrative Bureau
Changqing Petroleum Exploration Bureau
Sichuan Petroleum Administrative Bureau
North China Petroleum
Tuha Petroleum Exploration and Development Headquarters
Qinghai Petroleum Administrative Bureau
Jidong Exploration Corporation
......

Oil and gas businesses

Refining and chemical enterprises

CNPC Petroleum Exploration and Development Company
Fushun Petrochemical Corporation
Jilin Chemical Group Company
Liaoyang Petrochemical Fiber Company
Lanzhou Petrochemical
Urumqi General Petrochemical Works
Dalian Petrochemical Corporation
Jinzhou Petrochemical Corporation
Jinxi Refining and Chemical Complex
Linyuan Refinery
Qianguo Refinery   ......

Business performance

2009

CNPC crude oil production remained steady while that of natural gas continued to rapidly grow, with the year-round domestic output of crude oil being 103.13 million tons and that of natural gas being 68.3 billion m³.

In 2009, its comprehensive ranking was the fifth among the world's 50 largest petroleum corporations, and stayed at the 13th place in Fortune 500 rankings of the world's largest companies.

2010

Overseas and domestic petroleum and gas exploration and development of its group companies made a series of outstanding breakthroughs witnessing a continued rapid increase in crude oil and natural gas outputs, with the year-round domestic production of crude oil and that of natural gas being 105.41 million tons and 72.5 billion m^3 respectively.

In 2010, it still kept its fifth place in the comprehensive rankings of the world's 50 largest petroleum corporations, but improved its status as the 10th in Fortune 500 rankings of the world's largest companies.

CNPC crude oil output

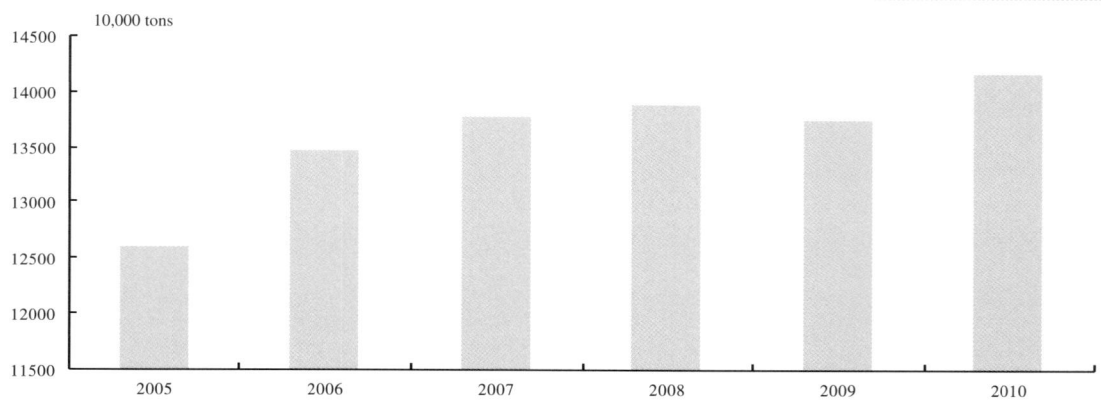

CNPC natural gas output

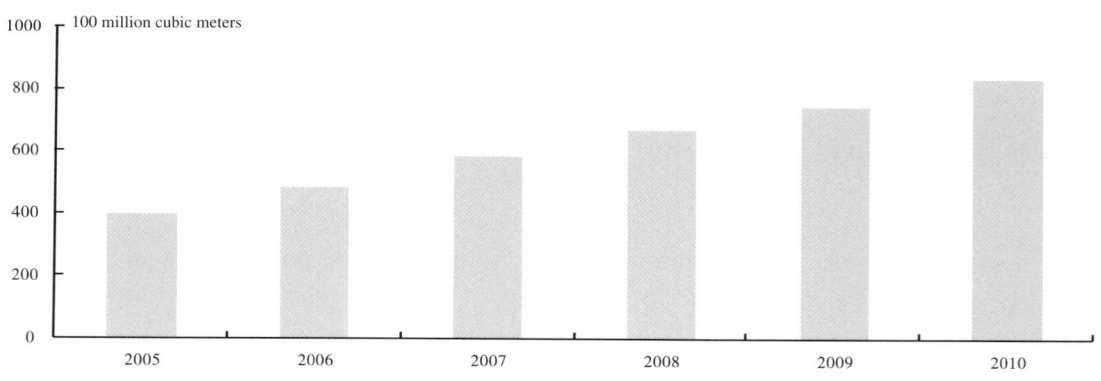

Business performance index

	2006	2007	2008	2009	2010
Total assets (100 million yuan)	14090	15990	18045	22214	26299
Owner's equity (100 million yuan)	8727	10358	11769	12695	13956
Sales revenue (100 million yuan)	8685	10007	12730	12183	17209
Newly added oil proven reserves (10,000 tons)	12333	12757	13907	13589	12878
Newly added natural gas proven reserves (100 million cubic meters)	1963	1496	1683	1540	1681
Total output of crude oil (including overseas equity yield) (10,000 tons)	13471	13762	13875	13745	14144
Total output of natural gas (including overseas equity yield) (10,000 tons)	480	578	664	738	829
Crude oil processing capacity (including overseas processing capacity) (10,000 tons)	12407	13088	13447	14082	16008
Processed oil output (including overseas production) (10,000 tons)	7869	8276	8512	9041	10185
Processed oil sales volume (domestic sales volume) (10,000 tons)	7522	8279	8293	8874	10247
Number of gas stations (domestic gas stations)	18207	18648	17456	17262	14996
Total length of crude oil pipeline (kilometers)	9816	12463	12931	13169	14807
Total length of processed oil pipeline (kilometers)	4311	4622	4610	8868	9257
Total length of natural gas pipeline (kilometers)	21138	22231	24225	28595	32801

Data source: Corporate Social Responsibility Report 2009

PetroChina company limited

Company overview

PetroChina Company Limited (abbreviated as "PetroChina"), a company limited by shares and founded in the process of CNPC restructuring on November 5, 1999, is the largest and leading producer and seller of petroleum and gas in China's oil and gas industry, and among the companies boasting the largest sales.

Its mainly runs the following businesses: exploration, development, production and marketing of crude oil and natural gas; refining of crude oil and oil products; production and marketing of basic and derivative chemicals and other chemical products; marketing and trading refined oil products; transportation of natural gas, crude oil and processed oil; and marketing of natural gas.

American Depository for Securities, H-shares and A-shares issued by the corporation were respectively listed on the New York Stock Exchange on April 6, 2000, on the Stock Exchange of Hong Kong Ltd. (Hong Kong Stock Exchange) on April 7, 2004, and on the Shanghai Stock Exchange on November 5, 2007.

Equity structure

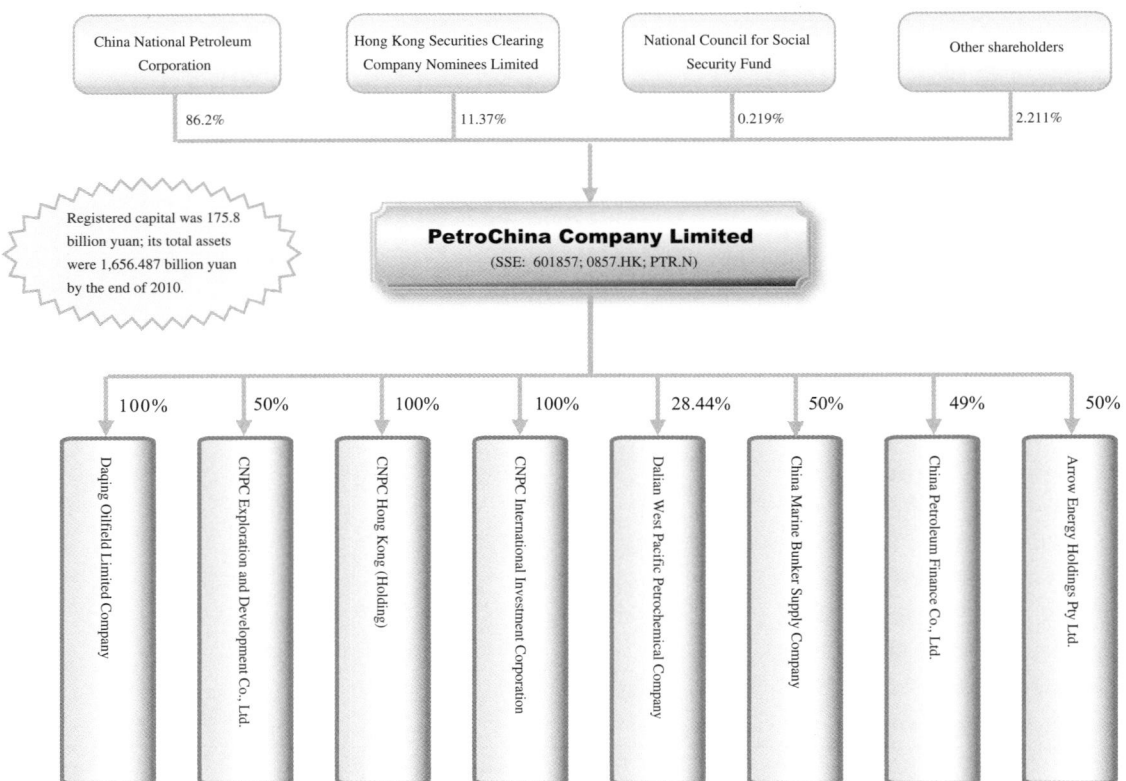

China National Petroleum Corporation — 86.2%

Hong Kong Securities Clearing Company Nominees Limited — 11.37%

National Council for Social Security Fund — 0.219%

Other shareholders — 2.211%

Registered capital was 175.8 billion yuan; its total assets were 1,656.487 billion yuan by the end of 2010.

PetroChina Company Limited
(SSE: 601857; 0857.HK; PTR.N)

- Daqing Oilfield Limited Company — 100%
- CNPC Exploration and Development Co., Ltd. — 50%
- CNPC Hong Kong (Holding) — 100%
- CNPC International Investment Corporation — 100%
- Dalian West Pacific Petrochemical Company — 28.44%
- China Marine Bunker Supply Company — 50%
- China Petroleum Finance Co., Ltd. — 49%
- Arrow Energy Holdings Pty Ltd. — 50%

Business performance

2009

The business performance went down, with a year-round operating revenue of 1019.27 billion yuan, down 5.0% year on year, and gaining a net profit of 103.17 billion yuan, down 9.3% year on year.

The oil and gas exploitation division achieved an operating profit of 105.02 billion yuan, reducing by 135.45 billion yuan; the company's realized price of crude oil was 53.90 dollars/barrel, falling by 38.4% year on year while that of natural gas was 0.814 yuan/m^3, up 0.1% year on year.

The refining and chemical division achieved an operating profit of 17.31 billion yuan, income increasing by 111.14 billion yuan, hence turning losses into profits; the year-round output of crude oil processing was 112.140 million tons, dropping by 2.5% year on year; gasoline production was 22.114 million tons, down 5.8% year on year; diesel production was 48.828 million tons, up 1.1% year on year; and ethylene production was 2.989 million tons, rising by 11.7% year on year.

The marketing division realized an operating profit of 13.26 billion yuan, increasing by 5.28 billion yuan, and its sales volume of processed oil hit 101.253 million tons, up 12.2% year on year.

The natural gas pipeline division achieved an operating profit of 19.05 billion yuan, increasing by 2.99 billion yuan, continuing to show a strong growing trend.

2010

Influenced by the rise in prices and sales volumes of crude oil, natural gas, etc., the whole-year operating revenue hit 1465.42 billion yuan, increasing by 43.8% year on year; it achieved a net profit of 139.87 billion yuan, up 35.6% year on year.

The oil and gas exploration division gained an operating profit of 153.70 billion yuan, with an increase of 48.68 billion yuan; the company's realized price of crude oil was 72.93 dollars/barrel, up 35.3% year on year, and that of natural gas was 0.955 yuan/m^3, up 17.3% year on year.

The refining and chemical division achieved an operating profit of 7.85 billion yuan, decreasing by 9.46 billion yuan; the company's whole-year volume of crude oil processing was 122.331 million tons, rising by 9.1% year on year; gasoline output hit 23.308 million tons, up 5.4% year on year; that of diesel was 53.745 million tons, up 10.1% year on year; and that of ethylene was 3.615 million tons, up 20.9% year on year.

The marketing division achieved an operating profit of 15.96 billion yuan, increasing by 2.69 billion yuan, and its sales volume of processed oil hit 120.833 million tons, up 19.3% year on year.

The Natural gas pipeline division achieved an operating profit of 20.42 billion yuan, increasing by 1.37 billion yuan.

China petroleum & chemical corporation

Company overview

China Petroleum & Chemical Corporation (CPCC), a Chinese corporation listed in Hong Kong, New York, London and Shanghai, and an energy and petrochemical corporation characteristic of a comprehensive integration of upstream, midstream and downstream industries, mainly runs the following businesses: oil and gas exploration and development, exploitation and marketing; natural gas pipeline transport; exploration, exploitation, pipiline transportation and sale of oil and gas, oil refining; production, marketing and storage of petrochemical fiber, chemical fertilizer and other petrochemical products. It is China's second largest producer of oil and natural gas.

In the 2009 Fortune rankings of the world's top 500 CPCC stepped forward into the top 10 for the very first time, ranking the ninth, ahead of the other Chinese enterprises in the rankings.

Equity structure

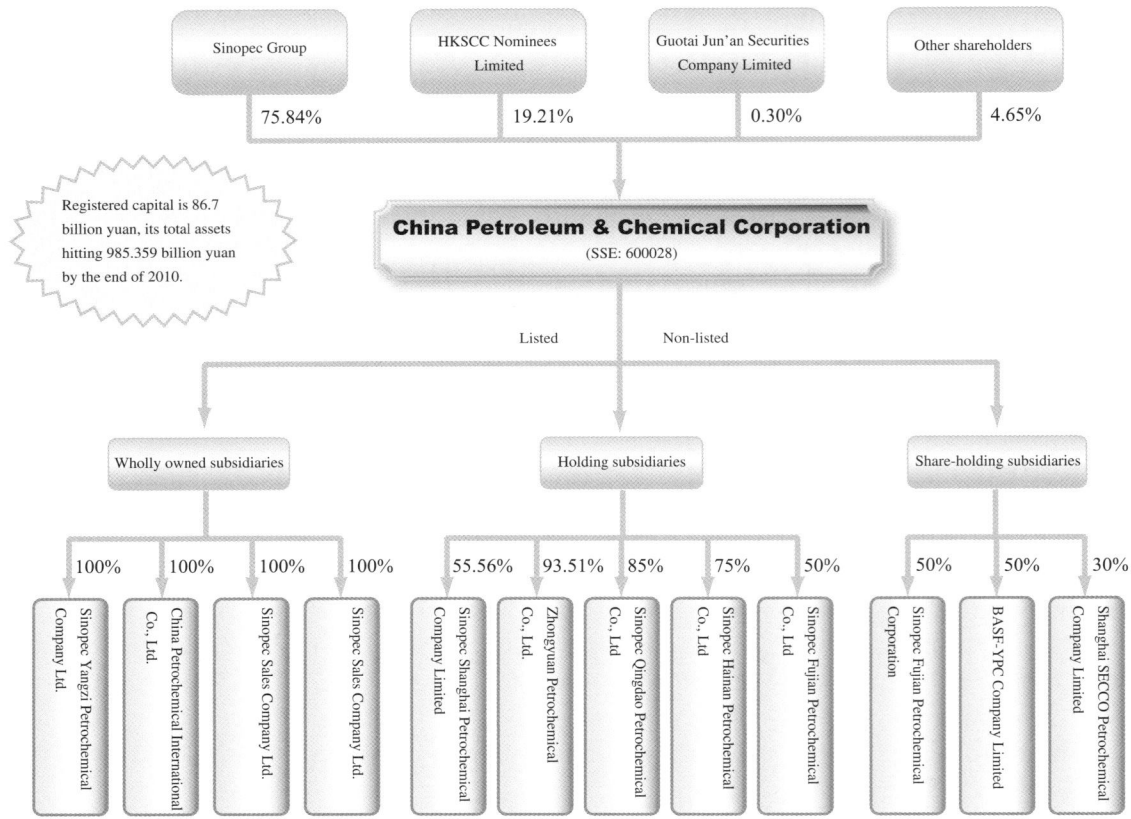

Business performance

2009

The corporation realized an operating revenue of 1345.05 billion yuan, down 6.9% year on year and a net profit of 61.29 billion yuan, up 115.5% year on year. The substantial increase in net profit was mainly due to the processed oil price reform.

For the exploration and development division, the average realized price of crude oil the year round was 2409 yuan/ton, dropping by 43.6% year on year; the turnover was 123.83 billion yuan, down 37.0% year on year; and the operating profit was 19.6 billion yuan, down 70.5% year on year.

For the oil refining division, the sluggish oil price and the pricing scheme reform of processed oil greatly improved its profitability. Its operating income fell by 19.2%, and the operating expenses dropped by 27.2%, thus realizing an operating profit of 23.1 billion yuan, with a substantial growth of 86.7 billion yuan year on year.

For the chemical division, since the unit cost of materials and price of chemical products experienced a substantial year-on-year fall, its year-round operating revenues were 218.46 billion yuan, down 13.8% year on year, and were its operating expenses 204.8 billion were yuan, down 23.1% year on year. Besides, it realized an operating income of 13.6 billion yuan, a year-on-year increase of 26.6 billion yuan.

For the marketing division, the domestic reforms of processed oil price and taxation narrowed the gap between prices overseas and at home. Along with this, the influence of sufficient market resources of processed oil and the intensified market competition all together led to the following results: the division's year-round turnover was 783.09 billion yuan, falling by 4.1% year on year; its operating expenses were 752.8 billion yuan, down 3.3% year on year; and it realized an operating profit of 30.3 billion yuan, down 21.3% year on year.

2010

Influenced by the relatively rapid growth of the domestic economy and the increase in the demand for oil and petrochemical products, the company achieved operating revenues of 1913.18 billion yuan, a year-on-year rise of 42.2%; meanwhile it realized a net profit of 70.71 billion yuan that belonged to the parent company, with a year-on-year rise of 12.8%.

Within the exploration and development division, the production of crude oil hit 46.090 million tons, up 0.1% year on year, with the realized price of crude oil being 3349 yuan/ton, a year-on-year rise of 45.4%; the production of natural gas was 12.50 billion m^3, up 47.6% year on year, with the average realized price being 1.155 yuan/m^3, rising by 23.8% year on year; the division realized an operating profit of 46.73 billion yuan, with a year-on-year increase of 22.58 yuan.

Within the oil refining division, the processing volume of crude oil was 211.130 million tons, up 13.2% year on year; the output of processed oil was 124.380 million tons, up 9.4% year on year; the division realized an operating profit of 14.87 billion yuan, falling by 12.604 billion yuan year on year.

Within the chemical division, the production of ethylene was 9.059 million tons, up 34.9% year on year; the division realized an operating profit of 14.76 billion yuan, with a year-on-year increase of 1.48 billion yuan.

For the marketing division, sales of processed oil hit 140.490 million tons, with a year-on-year rise of 13.3%; the division realized an operating profit of 30.62 billion yuan, rising by 340 million yuan year on year.

China national offshore oil corporation (CNOOC)

Company overview

China National Offshore Oil Corporation (CNOOC), a state-owned petroleum company founded in 1982, is China's third largest national petroleum corporation and China's largest offshore oil and gas producer. Till the present it has formed six big industry sectors, namely, oil and gas exploration and development, natural gas and electricity generation, oil refining and chemical fertilizers, professional technical services, comprehensive services and financial services. In the 2010 Fortune rankings of the world's top 500 it was listed at the 252nd place.

Upstream businesses

China National Offshore Oil Co., Ltd.

China Petroleum Research Center

Mid and downstream businesses

CNOOC Gas & Power Group

CNOOC Petrochemical Co., Ltd.

CNOOC Marketing Company

CNOOC Oil & Gas Development and Utilization Company

China Chemical Supply （Group） Company

China Blue Chemical Ltd.

China National Chemical Construction Corporation （CNCCC）

CNOOC Petrochemicals Import & Export Co., Ltd.

CNOOC Haixi Ningde Industrial Zone Development Co., Ltd.

Professional technical services

China Oilfield Services Limited (COSL)

Offshore Oil Engineering Co., Ltd.

CNOOC Energy Development Company Limited

China Offshore Oil Services Limited (Hong Kong)

Financial services

CNOOC Finance Co., Ltd.

Zhonghai Trust Co., Ltd.

CNOOC Insurance Limited (CIL)

CNOOC Investment Co., Ltd.

CNOOC Investment Holdings Co., Ltd.

Other businesses

CNOOC New Energy Investment Co., Ltd. (CNOOC-NEI)

China Offshore Oil Bohai Corporation

CNOOC Nanhai West Corporation

AEGON-CNOOC Life Insurance Co., Ltd.

CNOOC Nanhai East Corporation

CNOOC Donghai Corporation

CNOOC Industrial Company

CNOOC Infrastructure Company Pty Ltd.

CNCCC International Tendering Co., Ltd.

CNOOC Information Technology (Beijing) Ltd.

China Offshore Oil Press

Equity structure

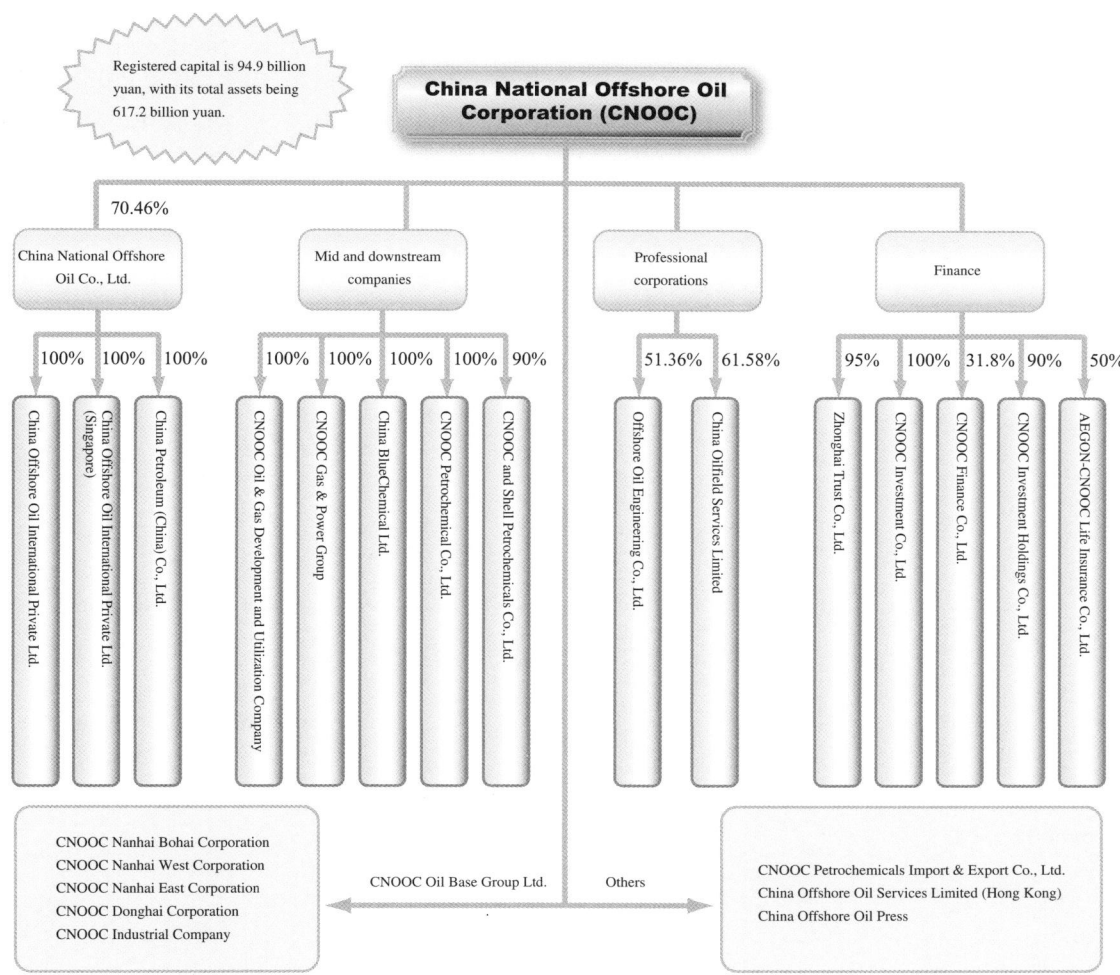

Registered capital is 94.9 billion yuan, with its total assets being 617.2 billion yuan.

China National Offshore Oil Corporation (CNOOC)

70.46%

China National Offshore Oil Co., Ltd.

Mid and downstream companies

Professional corporations

Finance

100%　100%　100%　　　100%　100%　100%　100%　90%　　　51.36%　61.58%　　　95%　100%　31.8%　90%　50%

China Offshore Oil International Private Ltd.

China Offshore Oil International Private Ltd. (Singapore)

China Petroleum (China) Co., Ltd.

CNOOC Oil & Gas Development and Utilization Company

CNOOC Gas & Power Group

China BlueChemical Ltd.

CNOOC Petrochemical Co., Ltd.

CNOOC and Shell Petrochemicals Co., Ltd.

Offshore Oil Engineering Co., Ltd.

China Oilfield Services Limited

Zhonghai Trust Co., Ltd.

CNOOC Investment Co., Ltd.

CNOOC Finance Co., Ltd.

CNOOC Investment Holdings Co., Ltd.

AEGON-CNOOC Life Insurance Co., Ltd.

CNOOC Nanhai Bohai Corporation
CNOOC Nanhai West Corporation
CNOOC Nanhai East Corporation
CNOOC Donghai Corporation
CNOOC Industrial Company

CNOOC Oil Base Group Ltd.　　　Others

CNOOC Petrochemicals Import & Export Co., Ltd.
China Offshore Oil Services Limited (Hong Kong)
China Offshore Oil Press

Business performance

2009

The company achieved a year-round operating revenue of 209.6 billion yuan, and a total profit of 52.4 billion yuan, down 22.7% year on year.

The outputs of major products witnessed a steady rapid growth, with year-round production of oil and gas hitting 47.66 million toe, and that of oil refining and chemical products amounting to 32 million tons.

2010

The company achieved a year-round operating revenue of 354.8 billion yuan, and a total profit of 97.7 billion yuan, up 86.5% year on year.

Oil and gas exploration made a breakthrough and developed in the Bohai Sea Gulf and the South Sea areas. The domestic oil and gas production witnessed a substantial growth, and also a steady growth of foreign oil and gas production, making a "double breakthrough" of 50 million tons domestically and 10 million tons overseas.

Company operating revenues and total profits

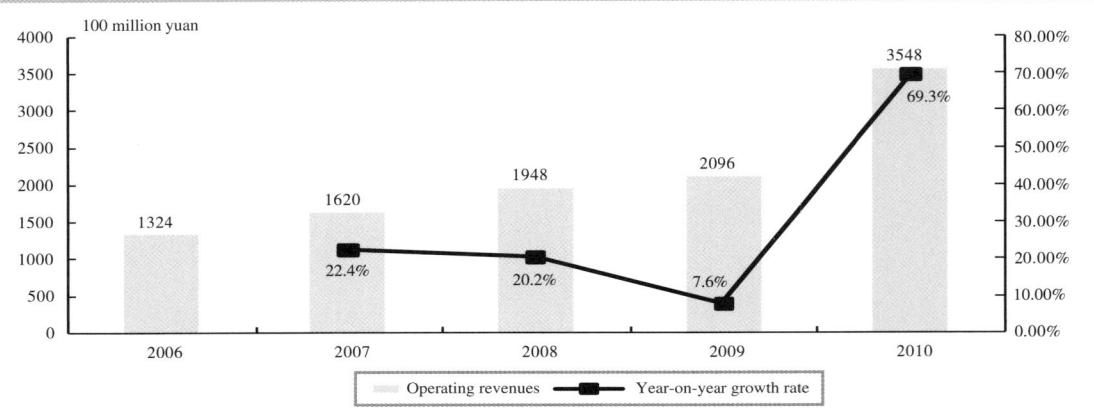

Data source: CNOOC Annual Report 2010

Total profits

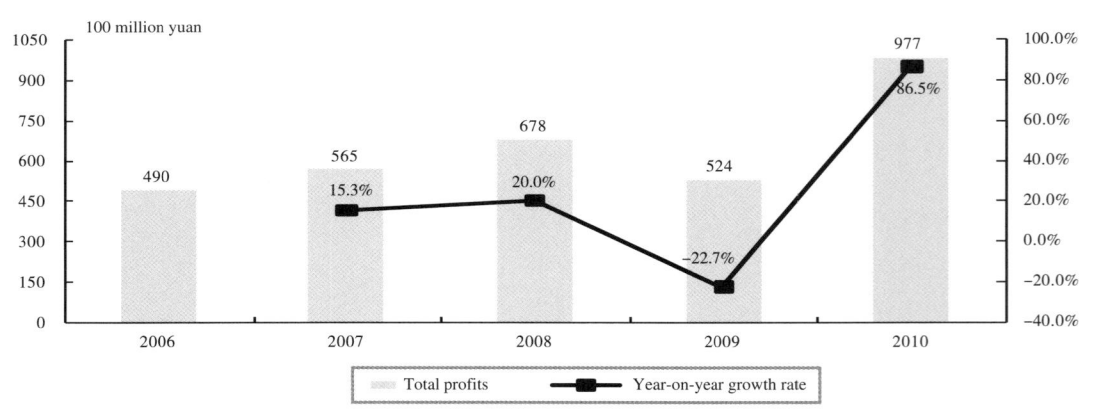

Data source: CNOOC Annual Report 2010

Outputs of major products

	Unit	2006	2007	2008	2009	2010
Upstream products	10,000 toe	4033	4047	4293	4766	6494
Domestic	%	86	85	85	82	80
Foreign	%	14	15	15	18	20
Petroleum liquid	10,000 toe	3154	3055	3244	3697	4958
Domestic	%	88	88	90	86	84
Foreign	%	12	12	10	14	16
Natural gas	100 million cubic meters	88	99	105	107	154
Domestic	%	79	75	73	70	66
Foreign	%	21	25	27	30	34
Downstream products	10,000 tons					
Urea	10,000 tons	193	185	195	190	201
Methanol	10,000 tons	27	71	77	81	87
Ethylene	10,000 tons	65	83	84	92	84
Asphalt	10,000 tons	177	147	120	415	791
Gasoline	10,000 tons				62	116
Kerosene	10,000 tons				77	105
Diesel	10,000 tons				277	463
Fuel oil	10,000 tons	626	658	569	341	475
Electricity generation	100 million kWh	35	52	60	86	149
LNG imports	10,000 tons	75	298	333	580	934

Data source: CNOOC Annual Report 2010

China national offshore oil corporation (CNOOC)

Company overview

China National Offshore Oil Corporation (CNOOC), a listed company registered and founded in Hong Kong in August 1999, is not only China's major producer of offshore crude oil and natural gas, but one of the world's largest independent corporation that deals with the development and production of crude oil and natural gas. It owns four major offshore oil production zones, i.e. The Bohai Sea Gulf, western South Sea, eastern South Sea and the East China Sea, its major businesses being as follows: exploration, development, production and marketing of oil and natural gas.

Equity structure

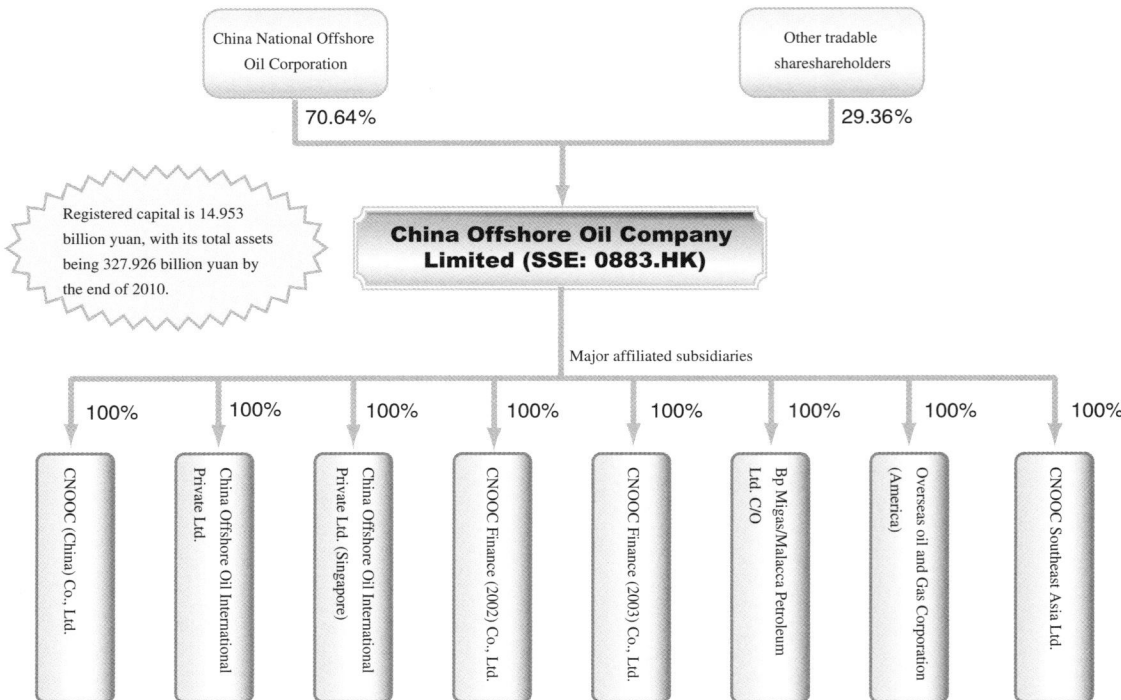

China National Offshore Oil Corporation

Other tradable shareshareholders

70.64%

29.36%

Registered capital is 14.953 billion yuan, with its total assets being 327.926 billion yuan by the end of 2010.

China Offshore Oil Company Limited (SSE: 0883.HK)

Major affiliated subsidiaries

100%   100%   100%   100%   100%   100%   100%   100%

CNOOC (China) Co., Ltd.

China Offshore Oil International Private Ltd.

China Offshore Oil International Private Ltd. (Singapore)

CNOOC Finance (2002) Co., Ltd.

CNOOC Finance (2003) Co., Ltd.

Bp Migas/Malacca Petroleum Ltd. C/O

Overseas oil and Gas Corporation (America)

CNOOC Southeast Asia Ltd.

Business performance

2009

The company achieved a total year-round revenue of 105.19 billion yuan, down 16.5% year on year; it gained a net profit of 29.49 billion yuan, down 33.6% year on year, which was mainly due to the fact that it takes the upstream business as its mainstay. When that was hit by a substantial decline in international oil prices, its performance experienced a downturn.

CNOOC's year-round net production of oil and gas achieved 227.7 million barrels of oil equivalent (boe), rising by 17.2% over the same period in the previous year. The annual average of daily net production hit 623,900 boe.

2010

The business performance enjoyed a large-scale growth in that the total year-round revenue hit 183.05 billion yuan, significantly rising by 74.0% year on year, along with a net profit of 54.41 billion yuan, up 84.5% year on year.

Thanks to the contributions made by the newly added production of oilfields that were put into operation since 2009, the excellent performance of the existing oil-gas fields, and the production of newly merged and acquired projects, the year-round net production of oil and gas hit 328.8 million boe, up 44.4% year on year. Besides, the total net production of oil and gas amounted to 901,000 boe/day.

CNOOC revenue and net profit

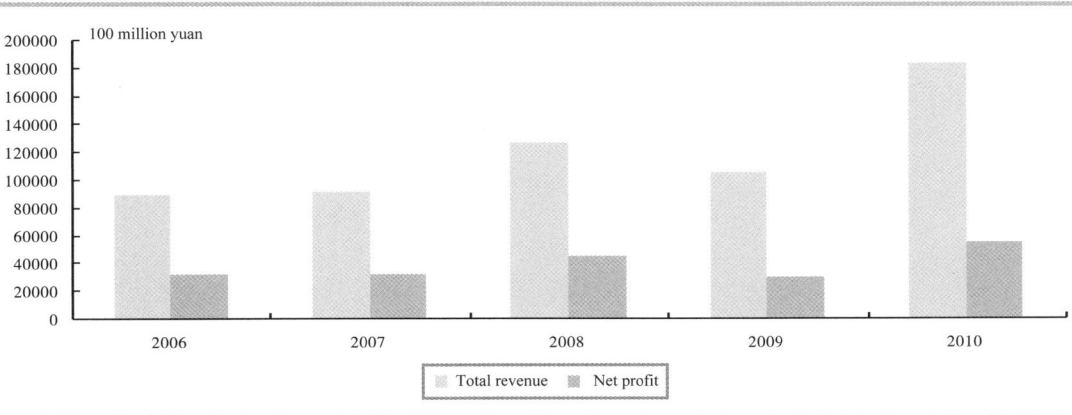

Data source: CNOOC Annals

CNOOC oil and gas net production

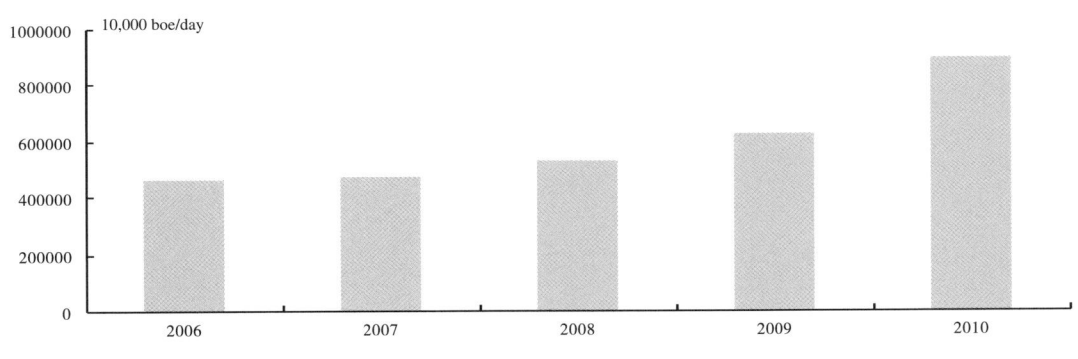

Data source: CNOOC Annals

Revenue structure in 2009   ## Revenue structure in 2010

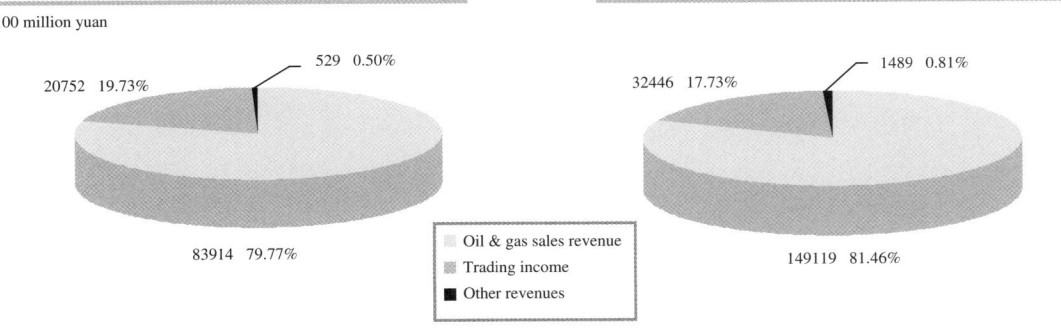

Data source: CNOOC Annals

Chapter 9   Coal Industry

Section 1: China's coal production and transport

Production

In 2009, China's cumulative output of raw coal was 2.97 billion tons, increasing by 170 million tons, up 6.1% year on year; in 2010, raw coal production maintained a continued rapid growth, with a cumulative year-round output of 3.24 billion tons, up 8.9% year on year.

China raw coal output

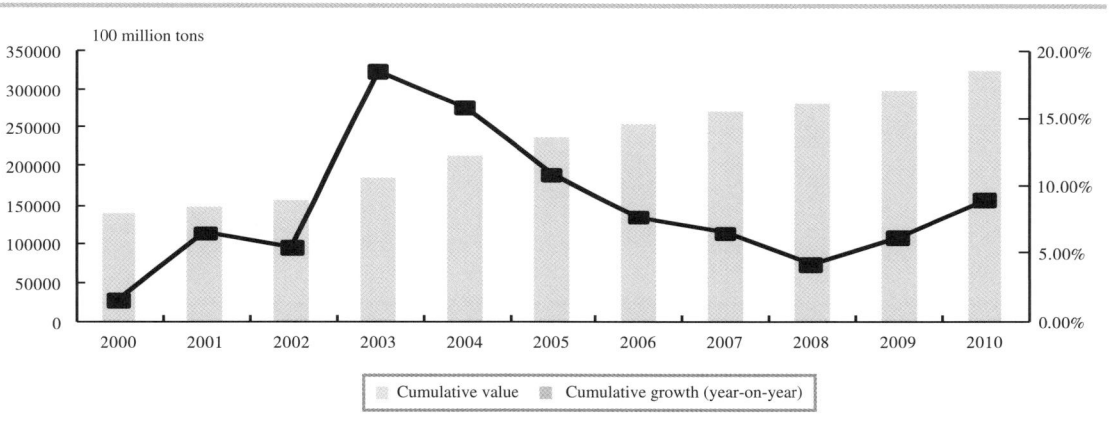

Data source: Statistical Yearbook

Looking at the outputs of different provinces and regions, in 2009 Inner Mongolia achieved 600.585 million tons of raw coal, up 19.6% year on year, surpassing that of Shanxi and ranking first in China in coal production at the provincial level. Shanxi Province, which used to rank first, achieved 593.540 million tons of raw coal, falling by 8.0% year on year.

In 2010, Inner Mongolia realized 786.647 million tons of raw coal, thus becoming the first region with raw coal output above 700 million tons, once again surpassing Shanxi. It indicates that the transfer of coal supply westward in China has been increasingly significant.

Raw coal outputs in different regions of China in 2009

10,000 tons

Region	2009	Year-on-year growth rate	Region	2009	Year-on-year growth rate
Beijing	641.3	10.8%	Henan	23018.1	8.0%
Hebei	8494.6	4.3%	Hubei	1058.5	−1.4%
Shanxi	59354.0	−8.0%	Hunan	6572.9	6.8%
Inner Mongolia	60058.5	19.6%	Guangxi	519.7	15.2%
Liaoning	6624.2	2.0%	Chongqing	4290.8	−8.1%
Jilin	4401.5	10.6%	Sichuan	8997.3	−5.2%
Heilongjiang	8748.7	−10.4%	Guizhou	13690.7	20.9%
Jiangsu	2397.4	−1.3%	Yunnan	5571.3	−30.6%
Zhejiang	13.2	0.6%	Shaanxi	29611.1	22.5%
Anhui	12848.6	10.3%	Gansu	3875.6	−3.6%
Fujian	2466.1	4.9%	Qinghai	1283.6	−0.8%
Jiangxi	2982.5	−9.7%	Ningxia	5509.5	27.4%
Shandong	14377.7	4.6%	Xinjiang	7646.0	13.5%

Data source: Energy Statistical Yearbook

Raw coal outputs in different regions of China in 2010

10,000 tons

Region	2009	Year-on-year growth rate	Region	2009	Year-on-year growth rate
Beijing	513.4	−19.9%	Henan	17908.8	−22.2%
Hebei	10199.3	20.1%	Hubei	1438.5	35.9%
Shanxi	74096.0	24.8%	Hunan	6200.0	−5.7%
Inner Mongolia	78664.7	31.0%	Guangxi	585.6	12.7%
Liaoning	5717.7	−13.7%	Chongqing	7659.7	78.5%
Jilin	4280.0	−2.8%	Sichuan	4377.1	−51.4%
Heilongjiang	9706.6	10.9%	Guizhou	15954.0	16.5%
Jiangsu	2160.9	−9.9%	Yunnan	9759.9	75.2%
Zhejiang	12.2	−7.6%	Shaanxi	36082.6	21.9%
Anhui	13144.8	2.3%	Gansu	4532.2	16.9%
Fujian	2092.1	−15.2%	Qinghai	1659.8	29.3%
Jiangxi	2746.2	−7.9%	Ningxia	6807.6	23.6%
Shandong	14892.0	3.6%	Xinjiang	10130.7	32.5%

Data source: Energy Statistical Yearbook

In 2009, the transfer of coal production in China westwards accelerated, with major newly added raw coal production concentrated in the northwestern region. The total newly added raw coal output of Inner Mongolia, Shaanxi, Ningxia and Xinjiang amounted to 225.893 million tons, accounting for 65.9% of the newly added output nationwide. In 2010, the trend of coal production being transferred towards the northwest continued.

In terms of different types of products, in 2009 the production of general bituminous coal and lignite witnessed a rapid growth, and their share of raw coal production nation-wide also went up rapidly. The main reason for the growth is that although in 2009 the raw coal output in Inner Mongolia saw a substantial growth, yet comparatively speaking it lacked quality anthracite and coking coal resources while abounding in lignite reserves.

Outputs of major coal types in 2009

10,000 tons

	Outputs of 2009	Year-on-year growth rate
Lignite	35435.1	20.6%
General bituminous coal	160762.8	17.9%
Anthracite	55334.6	15.1%
Coking coal	44941.5	5.1%

Data source: NBS

Transport

A characteristic of China's coal resources is that they are abundant in north and west but scarce in the south and east, with over 60% of coal resources distributed in the "three west" regions — Shanxi, Shaanxi and western Inner Mongolia. However, coal consumption is concentrated in the eastern coastal regions where the economy is relatively more developed than other regions of China. The inverse relationship between coal consumption and coal resources leads to the basic transport pattern of transporting coal from north to south and from west to east in China.

In China, coal transport mainly relies on railway, highway, coastal and inland water-ways, with railway as the primary means of coal transportation. In 2009, the rail freight volume of coal was 1.32 billion tons, slightly falling over the previous year. In 2010, the volume witnessed a rapid growth, achieving a year-round coal traffic volume of 1.55 billion tons, up 16.9% year on year.

Total volume of coal rail freight in China

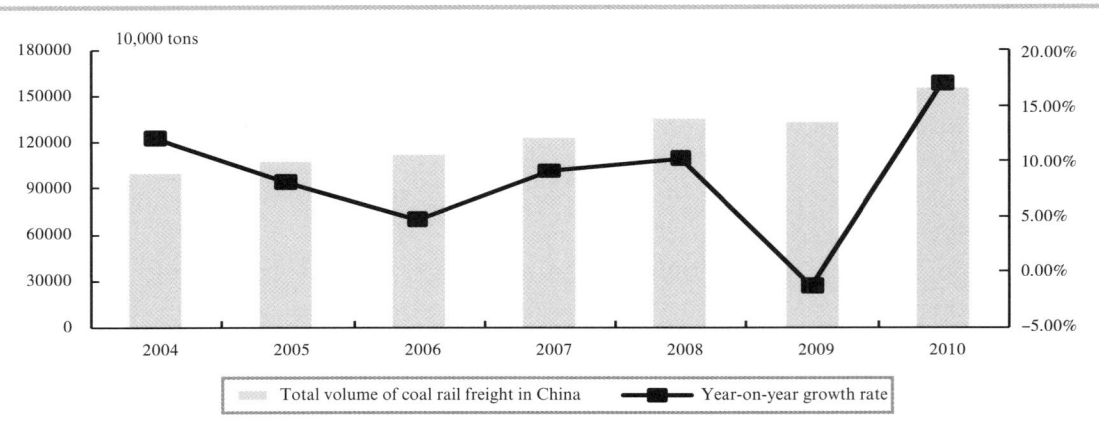

Data source: Wind

The coal rail transport system centers around Shanxi, Inner Mongolia and Henan, with its three west outbound railways concentrated along three major thoroughfares, namely, the north, the middle and the south. Its major coal railways are as follows:

North thoroughfare	Daqin Railway, Beijing-Yuanping Railway, Shenshuo-Shuohuang Railway, Fengsha Railway, Jitong Railway
Middle thoroughfare	Shitai Railway, Hanzhang Railway
South thoroughfare	Taijiao Railway, Longhai Railway, Ningxi Railway, Houyue Railway, Xikang Railway

Among the major thoroughfares of coal transport, in 2009 Daqin Railway achieved a cumulative traffic volume of 330.170 million tons, down 3.0% year on year; and that of Houyue Railway was 78.566 million tons, down 27.9% year on year.

In 2010, coal transport boasted a rapid growth among which Daqin Railway achieved a cumulative traffic volume of 405.040 million tons, up 22.7% year on year, and that of Houyue Railway was 88.762 million tons, up 13.0% year on year.

Coal traffic volume for each month in 2009

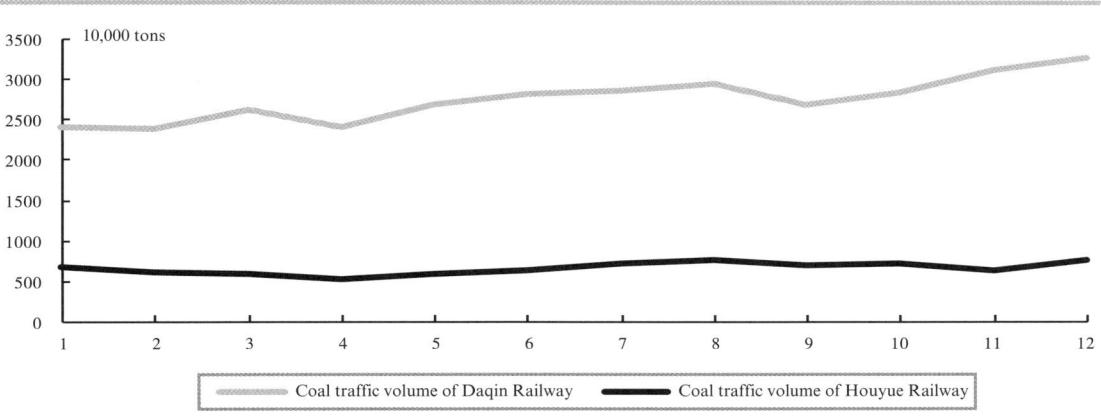

10,000 tons

Coal traffic volume of Daqin Railway　　Coal traffic volume of Houyue Railway

Data source: Wind

Coal traffic volume for each month in 2010

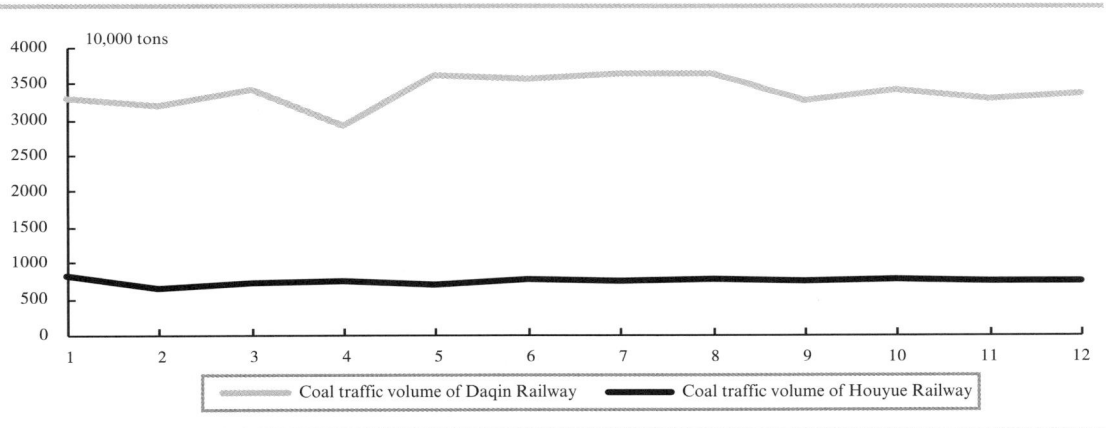

10,000 tons

Coal traffic volume of Daqin Railway　　Coal traffic volume of Houyue Railway

Data source: Wind

In 2009, the coal traffic volumes of major ports shrank, with their cumulative total traffic volume of coal being 461.469 million tons, falling by 52.544 million tons, down 10.2% year on year. Among this, the cumulative volume of domestic coal transported was 435.978 million tons, dropping by 31.301 tons, down 6.7% year on year, and that of foreign coal transported was 25.491 million tons, falling by 21.243 tons, down 45.5% year on year.

In 2010, the coal traffic volumes of major ports recovered their growth, with their year-round traffic volume of coal hitting 557.131 million tons, increasing by 95.662 million tons, up 20.7% year on year. Among this, the domestic coal traffic volume was 537.901 million tons, rising by 101.923 million tons, up 23.4% year on year; that of the foreign coal traffic volume was 19.232 million tons, falling by 6.259 million tons, down 24.6% year on year.

Coal traffic volume of major ports

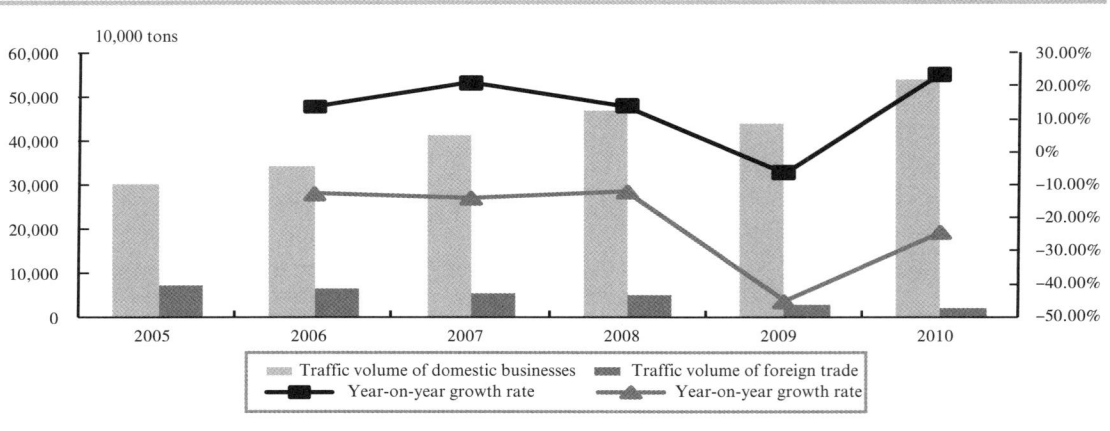

Data source: Wind

Section 2: Coal consumption and prices in China

In 2009, China coal consumption continued to increase, with total year-round consumption hitting 2.96 billion tons, increasing by 5.2%, with a slightly decreased growth rate; in 2010, China coal consumption maintained the momentum of growth, rising by 5.3%.

China coal consumption

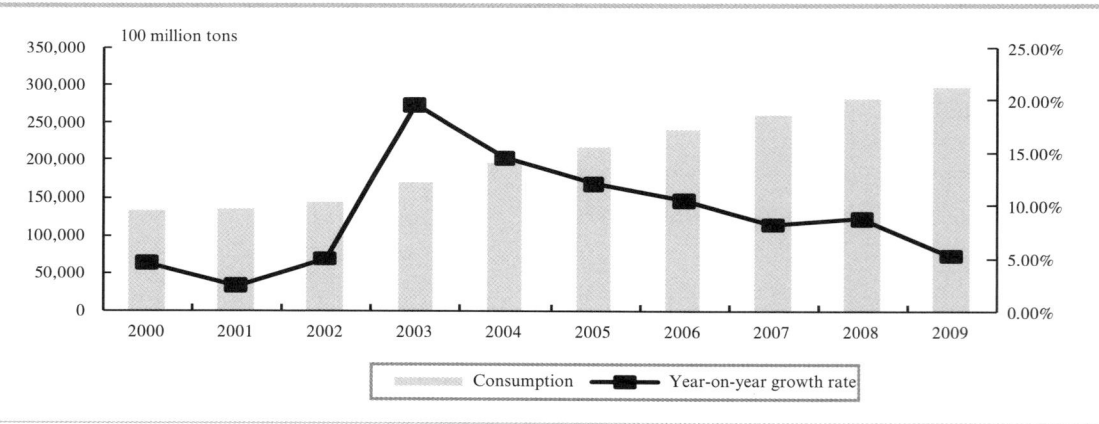

Data source: Statistical Yearbook

Looking at the trend for the whole year of 2009, demand for coal experienced a drastic change, from a slow recovery to a rapid growth, especially after the fourth quarter, during which time, as the national economic situation continued to turn for the better and the temperature in a majority of regions underwent a drastic fall, coal demand rapidly went up.

During the first half of 2011, pulled by the growth of high-energy-consuming industries and influenced by severe weather conditions, etc., coal supply and demand kept a tight balance; after midsummer arrived, as hydropower generation increased and the energy-saving and emission-reducing policies were executed, coal supply and demand were relatively loose; after winter arrived, coal supply and demand remained steady overall.

Looking at different consumption industries, coal is mainly utilized in the power industry, steel industry, building materials industry and chemical industry. In 2009, coal demand in the four largest industries shared over 80% of total coal demand; in 2010, due to the effects of a strengthened emphasis on the undertakings of saving energy and reducing emissions, coal consumption in these four industries underwent a declining growth rate.

Coal consumption of China's major coal-consuming industries (power, steel, building materials and chemical engineering)

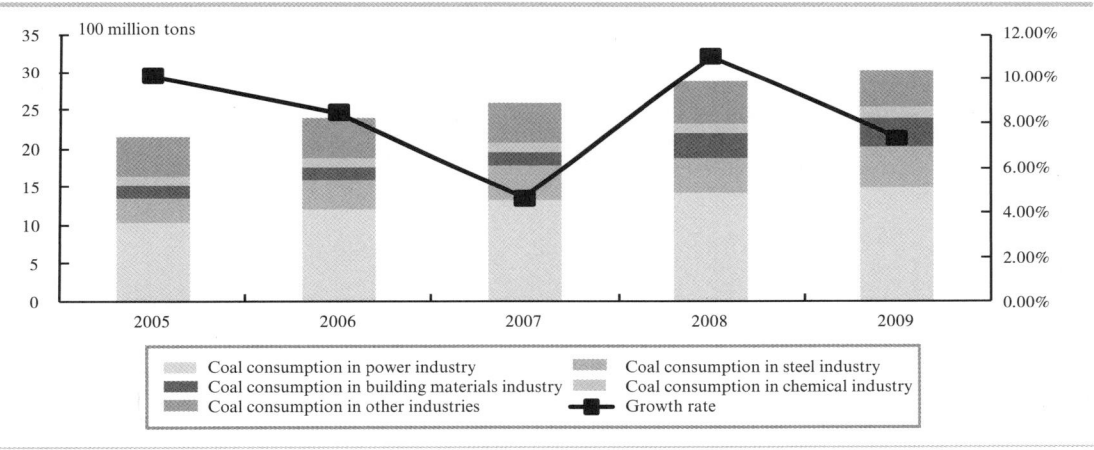

Data source: Statistical Yearbook

Coal prices

In 2009, prices of various types of coal in Qinhuangdao Port that normally function as the indicator for the industry witnessed a fluctuating adjustment: in the first half of the year, having undergone a drastic decline during the second half of 2008, coal prices remained basically steady; in 2010, coal prices witnessed periodical fluctuations with the changes in coal supply and demand, showing a W-shaped trend.

In terms of the year-round trend, in early 2009, as the coal inventory level of Qinhuangdao Port went up step by step, prices of various types of coal in this port also experienced decline to different degrees. However, since the last third of October, as heavy industry was taking a gradual turn for the better and the growth of electricity consumption accelerated, along with the coming of the peak of coal utilization in winter and the difficulty in transport caused by the heavy snow in North China, steam coal prices started to witness a trend of continued growth. In November, with the peak of coal utilization in winter, prices underwent a new round of rapid growth, hitting a new high for the year.

In early 2010, affected by the extreme cold weather, the flourishing coal demand caused prices to remain at a relatively high level; however, with the gradual fall in demand, coal prices reached the lowest level of the year in the last third of March. During the second quarter, along with the continued recovery of production was the rebound in coal prices; during the summer peak, due to sufficient coal inventory, prices underwent a fallback; then in the last third of October, influenced by the increase in energy prices and the ensuing storage of coal for the winter, coal prices witnessed a rapid increase.

Coal prices during the years 2009–2010

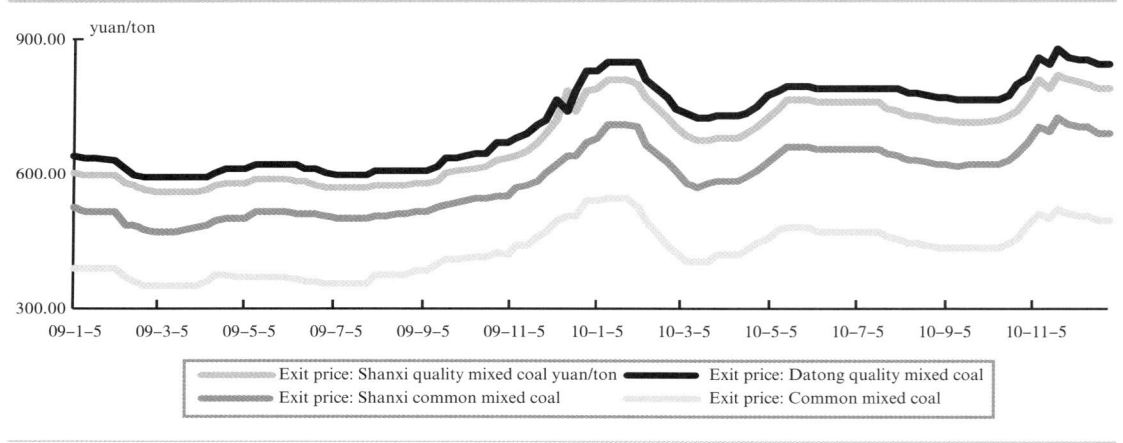

Data source: CCTD

Section 3: Analysis of coal imports and exports

Influenced by the global financial crisis, world energy demand witnessed a rapid decline, leading to a fallback of international coal prices at a high level. By contrast, domestic coal prices remained steady at a high level, thus creating the inversion between foreign and domestic prices. Under this influence, in 2009 coal imports experienced a substantial growth, while coal exports underwent rapid shrinking. In this year China became for the first time a net coal importer.

In 2009, China's cumulative coal imports hit 125.38 million tons, up 211.9%; it exported 22.40 million tons of coal, down 50.7% year on year; its net coal imports were 102.98 million tons, completing its change from net coal exporter to net coal importer.

In 2010, China coal imports continued to maintain the momentum of a rapid growth and recorded a new high; the year-round cumulative coal imports were 164.78 million tons, up 30.9% year on year; in contrast, coal exports continued the downward trend during the recent years, with year-round cumulative exports being 19.03 million tons, down 15% year on year, reaching a record low since 1991.

Coal imports and exports in China

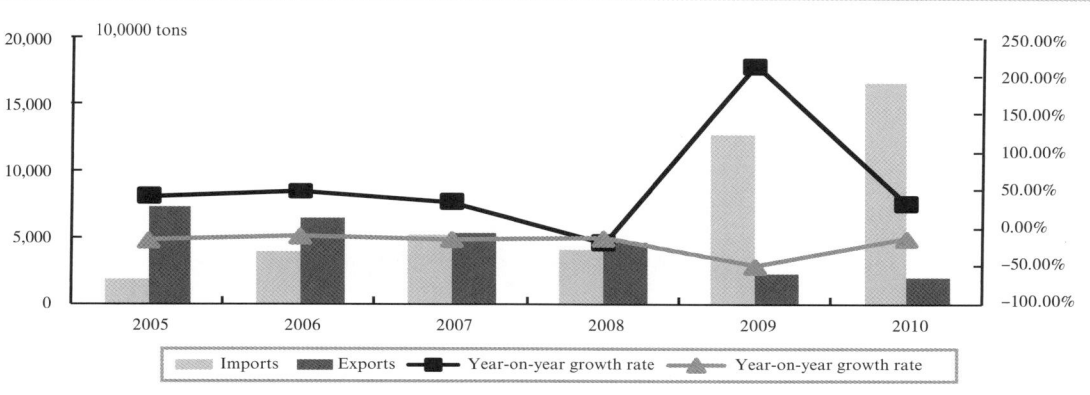

Data source: General Administration of Customs (China)

Looking at the trend for the whole year, in 2009 our country witnessed a drastic increase in coal imports; the imports for the month November hit 16.38 million tons, increasing by 5.1 times year on year, with a sequential growth rate of 29.3%. In contrast, coal exports fluctuated around a low level. In November coal exports were 2.07 million tons, down 53.7% year on year, with a sequential growth rate of 44.1%; however, in 2010 coal imports tended to remain steady while coal exports showed a gradually declining tendency.

China's monthly coal imports/exports in 2009

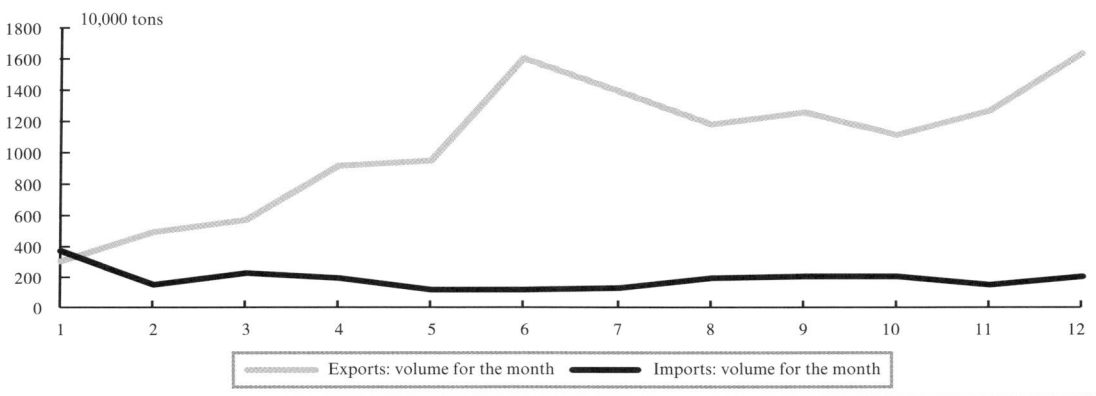

Data source: General Administration of Customs (China)

China's monthly coal imports/exports in 2010

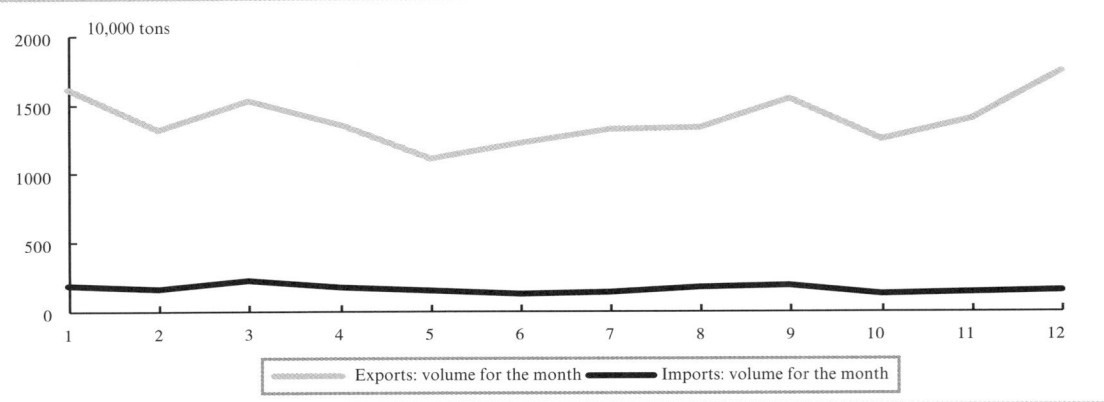

Data source: General Administration of Customs (China)

In terms of export regions, coal exports were mainly concentrated in Korea, Japan and Taiwan.

In 2009, China exported 9.88 million tons of coal to Korea, falling by 40.3%; it exported 6.40 million tons of coal to Japan, falling by 52.1%; and it exported 4.93 million tons of coal to Taiwan, down 53.5%. The sum of coal exports to these three regions accounted for 94.7% of the total within the same period.

In 2010, China exported 7.24 million tons of coal to Korea, falling by 26.7%; 6.47 million tons of coal to Japan, slightly rising by 1.1%; and 4.43 million tons of coal to Taiwan, down 10.2%. The sum of coal exports to these three regions accounted for 95.3% of the total within the same period.

Regional composition of coal importers from China in 2009

Regional composition of coal importers from China in 2010

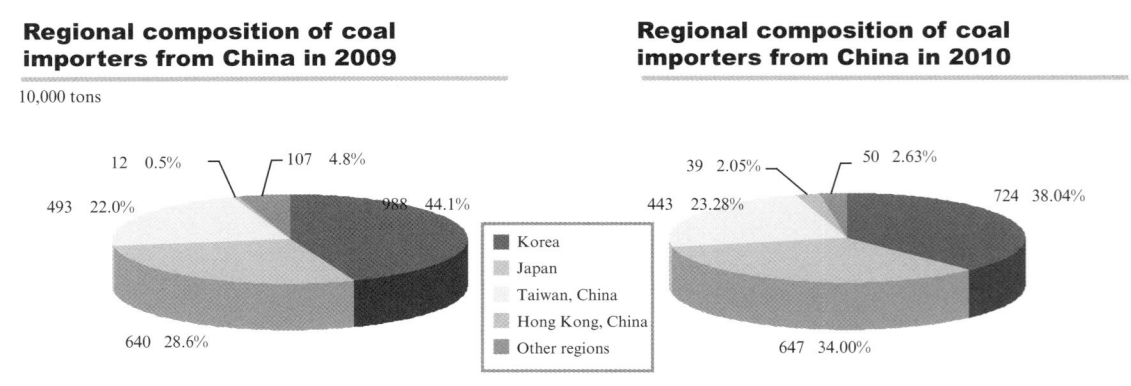

Data source: General Administration of Customs (China)

In terms of import regions, coal import channels witnessed an increase, with coal sources further dispersed.

In 2009, coal imports from ASEAN, Australia and Russia saw a significant growth: 54.56 million tons from ASEAN, up 91.2%; 43.95 million tons from Australia, rising by 11.4 times; 11.78 million tons from Russia, increasing by 14.5 times. The sum of coal imports from the three regions made up 87.7% of the total imports in the same period.

In 2010, imports from Indonesia amounted to 55.03 million tons, a substantial growth year on year, making it China's largest coal importer, closely followed by Australia, Vietnam, Mongolia and Russia. The total coal imports from these five countries accounted for 84% of the total imports nationwide.

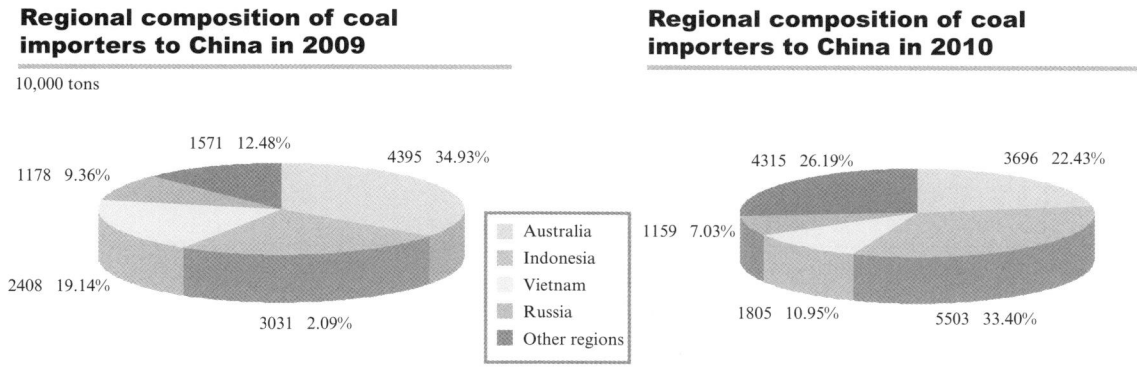

Regional composition of coal importers to China in 2009

10,000 tons

Regional composition of coal importers to China in 2010

Legend:
- Australia
- Indonesia
- Vietnam
- Russia
- Other regions

Data source: General Administration of Customs (China)

Looking at the types of coal imported and exported, in 2009 bituminous coal was the major type exported, and coking coal boasted the highest growth rate of import. For the whole year, China exported 19.11 million tons of bituminous coal, down 51.3%, making up 85.3% of total coal exports in the same period. Among this, coking coal exports were 636,000 tons, down 81.6%; exports of other bituminous coal were 18.48 million tons, down 48.3%. Within the same period, China imported 72.46 million tons of bituminous coal, rising by 3.4 times and making up 57.6% of total coal imports in the same period. Among this, coking coal imports were 34.42 million tons, up 4 times; those of other bituminous coal were 38.03 million tons, up 2.9 times. In addition, anthracite imports hit 34.33 million tons, rising by 77.1% with a share of 27.3%.

In 2010, the major type of coal exported still was steam coal, yet its exports witnessed a gradual fallback. In contrast, steam coal and the other types boasted the largest amount of imports, hitting 82.52 million tons, up 71.4% year on year; coking coal imports were 41.82 million tons, up 34.9% year on year; and anthracite imports fell to 23.44 million tons, down 24.7% year on year.

Section 4: Economic benefits and competition pattern of China's coal industry

In the fourth quarter of 2009, the booming index of China's coal industry read 99.79 points (100 points in 2001), experiencing a reversing trend. A rapid recovery of downstream businesses laid a solid foundation for the recovered boom of China's coal industry.

In the fourth quarter of 2010, the booming index of China's coal industry indicated 104.61 points (100 points in 2001), changing from rise to decline. Due to the influence of energy saving and emission reduction, the booming index that had been consecutively rising for four quarters began to fall, and this declining tendency is expected to continue.

Booming index of China's coal industry during the years 2002–2010

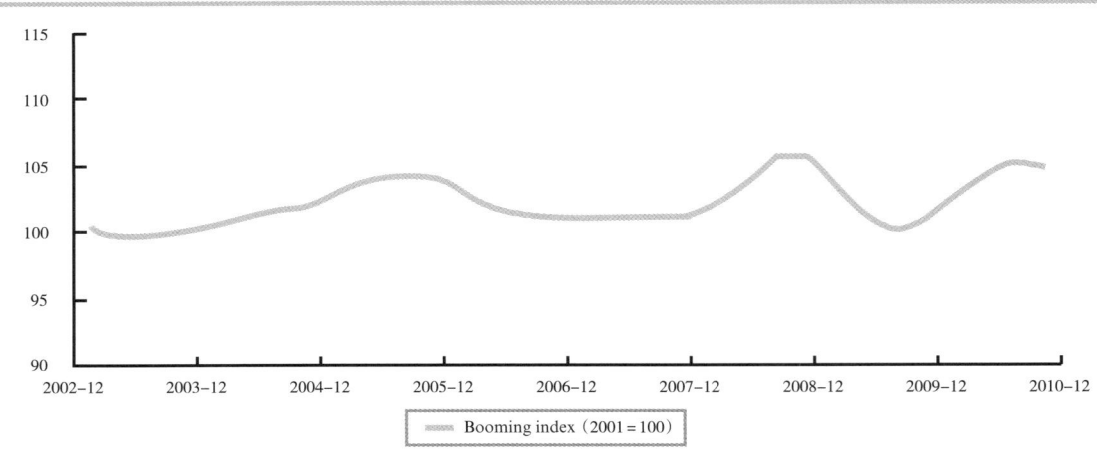

Booming index (2001 = 100)

Data source: Center for China Economy Industry Boom Index Research

In the fourth quarter of 2010, 7 of 10 indexes making up the early warning index for China's coal industry were situated within the "green light range", one in the "red light range" and one in the "pale blue light range". In the fourth quarter of 2010, the early warning index changed into 32 points, from the "pale blue light range" to the "green light range" and then rushing to the "yellow light range", thus realizing a fallback for the first time.

Booming index lights

Index	2007	2008				2009				2010			
	12	3	6	9	12	3	6	9	12	3	6	9	12
Total-profit growth rate													
Total-tax growth rate													
Annual average rate of employee development													
Producer's price index													
Growth rate of crude oil production													
Inventory turnover ratio													
Receivables turnover ratio													
Rate of sales income growth													
Rate of fixed investment growth													
Exports growth rate													
Early warning index	30.0	32.0	39.0	41.0	37.0	28.0	23.0	21.0	27.0	33.0	35.0	36.0	32.0

Source: Center for China Economy Industry Boom Index Research

Note: The early warning light describe the development of an industry concerned ⬤ signals "too fast" ("excessive growth"), ⬤ signals "fast", ⬤ signals "normal and stable", ⬤ signals "slow", and ⬤ signals "too slow". Besides, it also assigns to each index light a different score, all of which form the comprehensive early warning index, also indicated by the five kinds of light , with the same above.

In 2009, the fixed investment delivery of China's coal mining and dressing industry hit 302.15 billion yuan, up 25.9% year on year, maintaining a substantial growth; in 2010, this delivery changed into 37.70.3 billion yuan, up 23.3% year on year, still keeping a relatively rapid growth.

Fixed investment delivery of China's coal mining and dressing industry

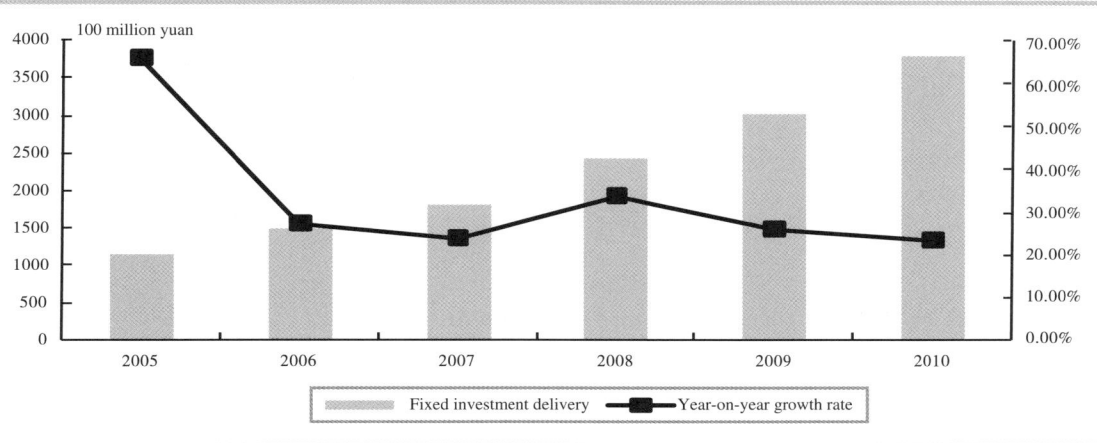

Data source: NBS

In 2009, the sales revenue of the coal industry witnessed an overall growth: during the months January to November, the total sales revenue was 1541.97 billion yuan, the growth rate rising by 10.4% year on year, which was mainly attributed to a year-on-year growth in coal output.

In the period from January to November of 2010, the coal industry achieved up to 2115.50 billion yuan, a substantial year-on-year growth of 41.8%. However, seen from a whole-year perspective, the rate of sales revenue growth gradually became stable after going up.

Revenues of major businesses of China's coal mining and dressing industry

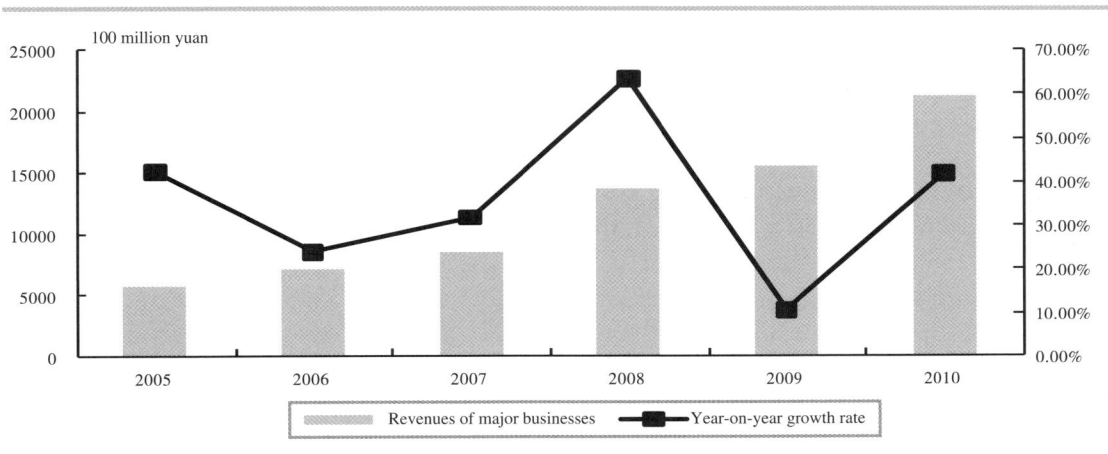

Data source: NBS

In 2009, the average selling price of crude coal dressing of large-scale coal enterprises in China was 427.1 yuan/ton, falling by 64.8 yuan/ton year on year, down 8.9%; the unit cost of crude coal dressing was 389.5 yuan/ton, rising by 4.3 yuan/ton year on year, up 1.1%, thus leading to an increase in enterprise cost.

In the period from January to November of 2009, the coal industry achieved a profit of 182.7 billion yuan, down 9.4% year on year. What's worthy of attention is that although the coal industry witnessed an increase in year-round coal output and business revenues, yet due to a fall in the average coal price, profits saw a drop.

In 2010, benefiting from the adjustment in high-level fluctuations of coal prices, the profitability of the coal industry continued to grow. During the months January to November, the coal industry achieved a profit of 293.02 billion yuan, substantially rising by 61.1% year on year.

Total profits of China's coal mining and dressing industry

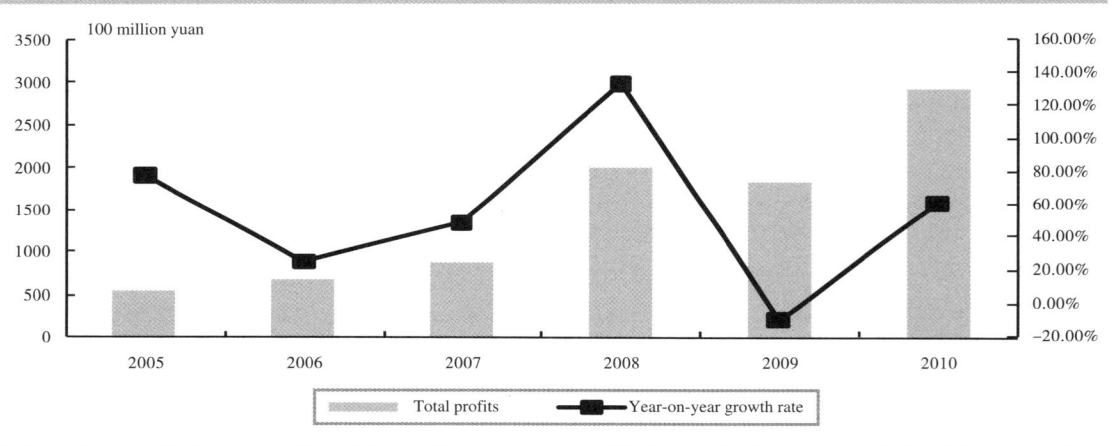

Data source: NBS

In 2009, the tax payable of the national large-sized coal industry totaled 146.71 billion yuan, up 15.5% year on year, among which value added tax (VAT) payable was 74.98 billion yuan, up 14.2% year on year. This was mainly due to the fact that in 2009 the VAT rate of the coal industry was adjusted from 13% up to 17%, and that policies on expanding domestic demand promoted the simultaneous growth of coal production and sales, both of which led to a steady increase in the tax of the coal industry. In 2010 the total tax continued to go up at a significantly increased growth rate.

Coal mining and dressing industry: cumulative value of total tax

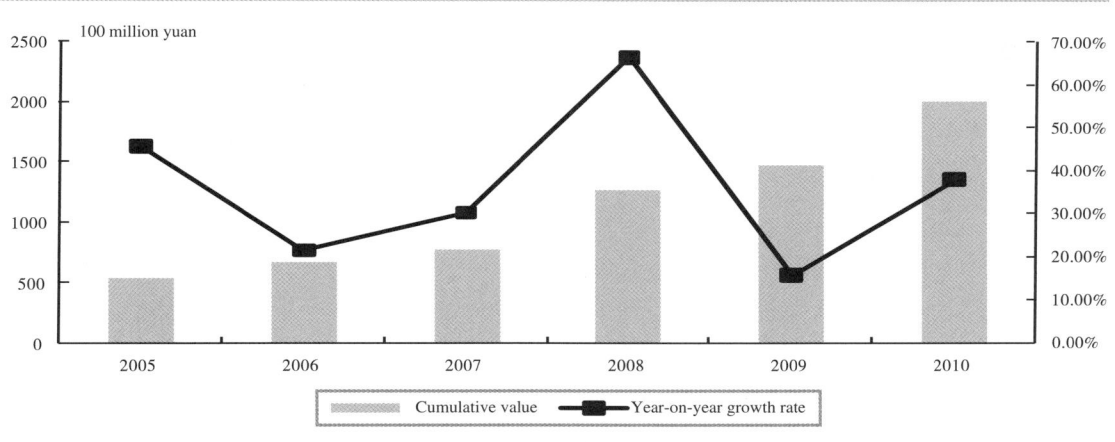

Data source: NBS

In 2009, the coal industry put emphasis on promoting integration of coal resources and acquisition and restructuring of the coal industry, and witnessed a smooth development in scale. The number of coal mines nationwide declined significantly, which thus led to an increase in the industry's concentration ratio.

Number of coal mines nationwide

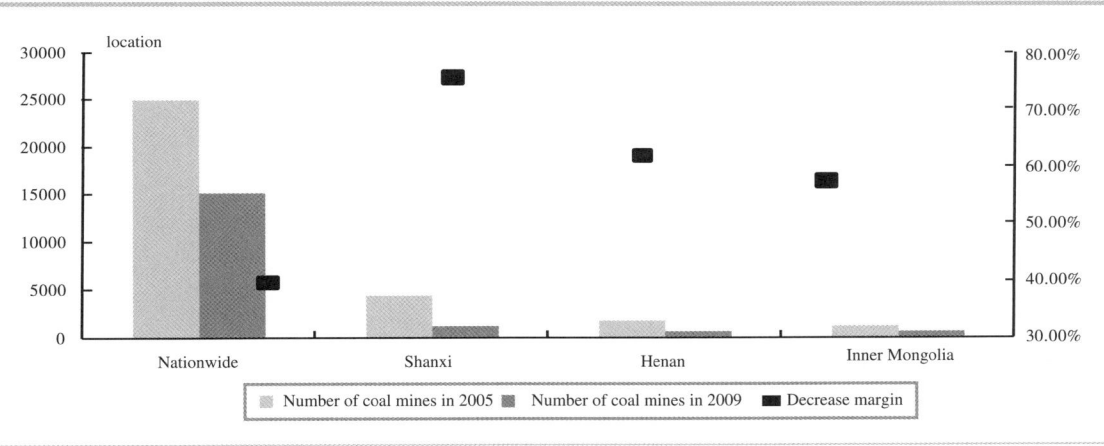

Data source: China National Coal Association

Section 5: Typical coal enterprises in China

China Shenhua Energy Company Limited

Company overview

China Shenhua Energy Company Limited, a mega energy enterprise, characterized by its integration of the power, railway, port, shipping, coal-to-liquid and coal chemical industries, as well as its integration of production, transport and marketing, is China's largest coal enterprise in scale and degree of modernization and the world's largest coal franchiser, ranking among the world's top 500.

Equity structure

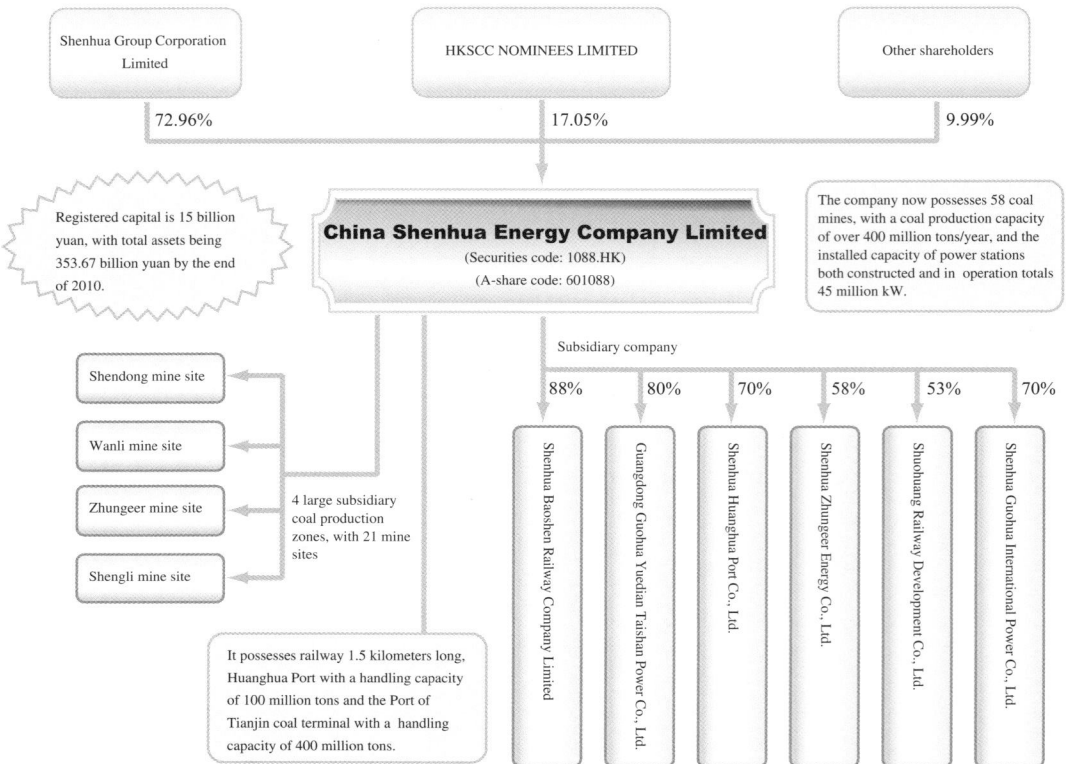

Business performance

2009

The company realized a year-round operating revenue of 121.31 billion yuan, up 13.2% year on year, and a net profit of 30.28 billion yuan, up 16.6% year on year.

The company achieved total year-round coal production of 210 million tons, and a sales volume of commercial coal of 250 million tons, up 13.2% and 9.3% respectively, realizing steady growth of coal production and sales.

The total year-round electricity generation was 105.09 billion kWh, and electricity sales hit 97.72 billion kWh, rising by 7.5% and 8.2% respectively.

2010

The company realized a year-round operating revenue of 152.06 billion yuan, up 25.3% year on year, which was mainly due to the simultaneous increase in coal production and price as well as electricity sales. It achieved a whole-year net profit of 42.51 billion yuan, up 22.3% year on year.

The company realized year-round coal production of 220 million tons, up 6.9% year on year, which was mainly due to the increase in coal production of Haerwusu mine, Heidaigou mine, and Jinjie mine, etc., as well as production growth brought about by tender offers or assets purchases. It achieved a year-round coal sales volume of 290 million tons, up 15.1% year on year.

The total all-year electricity generation and sales volume were respectively 141.15 billion kWh and 131.41 billion kWh, up 34.3% and 34.5% year on year.

YanZhou Coal Mining Co., Ltd.

Company overview

YanZhou Coal Mining Co., Ltd., founded in Zoucheng City, Guangdong Province, mainly deals with exploitation, dressing and processing, and marketing of underground coal and coal railway transport. Established in 1997, it was listed in 1998 in three places, i.e. Hong Kong, New York and Shanghai.

Equity structure

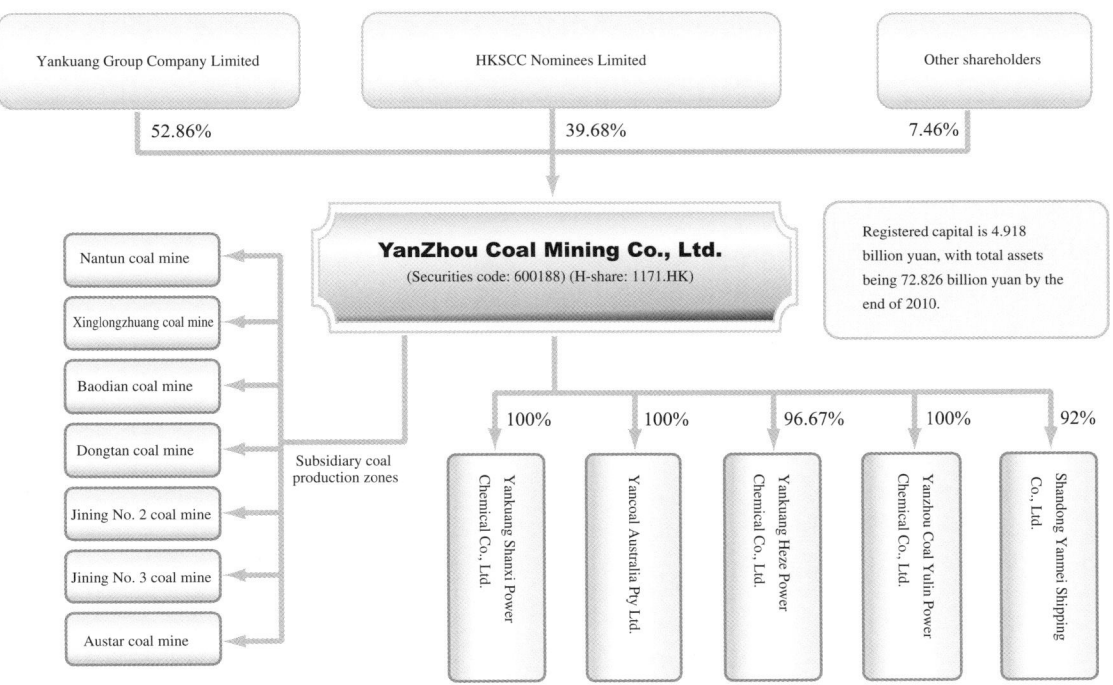

Business performance

2009

The company realized a year-round operating revenue of 21.50 billion yuan, down 18.4% year on year; it achieved an all-year operating profit of 5.41 billion yuan, down 37.5% year on year; and it achieved a net profit of 3.88 billion yuan that belonged to the parent company shareholder, down 38.6% year on year.

For the whole year the company achieved 35.77 million tons of commercial coal production, up 0.7% year on year, and 38.02 million tons of coal sales, up 1.2% year on year.

In 2009, the average selling price of coal was 529.0 yuan/ton, falling by 20.0% over the previous year in the same period, which was the major factor causing the decline of the company's business performance for the year 2009.

2010

The company realized a year-round operating revenue of 34.84 billion yuan, up 62.1% year on year, one part of the revenue growth benefiting from the increase in commercial coal sales, and the other part from the year-on-year rise in the average selling price of commercial coal in 2009. The total profit was 12.11 billion yuan, up 123.9% year on year, among which the net profit attributable to the shareholders of listed companies was 9.01 billion yuan, up 132.16% year on year.

Within the year, coal production and sales both witnessed a substantial growth: raw coal production hit 49.40 million tons, up 36.1% year on year, among which Yancoal Australia Pty Ltd. was the primary contributor to the growth of commercial coal outputs; the sales volume of commercial coal amounted to 49.63 million tons (including 5.38 million tons of bought-in coal), up 30.6% year on year.

In 2010, the comprehensive selling price of coal was 663.4 yuan/ton, significantly rising by 25.4% year on year, approximate to the record high, namely, 663.9 yuan/ton in the year 2008.

Shanxi Guo Yang New Energy Co., Ltd.

Company overview

Shanxi Guo Yang New Energy Co., Ltd., situated in Yangquan, Shanxi Province, one of China's six largest anthracite bases, and founded in December 1999, was listed in August 2003 with public offerings. It mainly deals with exploitation, dressing and processing, and marketing of coal, and generation and marketing of electricity.

Equity structure

Yangquan Coal Industry (Group) Co., Ltd.

Shanghai Pudong Development Bank-GF Securities Growth Stock Securities Investment Fund

Other shareholders

58.34%

1.12%

40.54%

Shanxi Guo Yang New Energy Co., Ltd.
(Securities code: 600348)

Registered capital is 481 million yuan, with total assets being 22.258 billion yuan by the end of 2010.

No. 1 Mine of Yangquan Coal Group

Subsidiary coal production mine

No. 2 Mine of Yangquan Coal Group

100%

100%

Yangquan Coal Group Coal Washing Branch

Yangquan Coal Group Power Generation and Supply Branch

Business performance

2009

The company achieved a year-round operating revenue of 20.01 billion yuan, up 17.5% year on year, and a total profit of 2.49 billion yuan, up 108.1% year on year. The net profit attributable to listed company shareholders was 1.86 billion yuan, rising substantially by 91.8% year on year.

The whole-year raw coal production totaled 20.96 million tons, up 18.2% year on year, which was mainly accounted for by the fact that the company acquired Xinjing Coal Mine from its parent company in the year; it achieved a sales volume of 39.53 million tons, up 8.8% year on year.

In 2009, the company's comprehensive selling price of coal was 439.6 yuan/ton.

2010

The company achieved a year-round operating revenue of 27.941 billion yuan, 39.67% up year on year, with 2.413 billion yuan of net profit attributable to the parent company, which rose by 29.96% year on year.

Within the year, raw coal production reached 26.21 million tons, up 25.01% year on year, the main reason for this growth being the acquisition of Xinjing Coal Mine from its parent company. The year-round coal sales hit 45.38 million tons, up 14.8% year on year, the main reason for this growth being the increase in the amount of domestic coal production and bought-in coal.

In 2010, the comprehensive selling price of coal was 489.4 yuan/ton, up 11.3% year on year. Among this, the prices of washed lump coal dust powder and washed coal dust were 731.1 yuan/ton, 835.6 yuan/ton and 405.9 yuan/ton, respectively, up 0.0%, 27.6% and 10.4% year on year respectively.

Henan Shenhuo Coal & Power Co., Ltd.

Company overview

Henan Shenhuo Coal & Power Co., Ltd., listed on the Shenzhen Stock Exchange on August 31, 1999, mainly deals with coal exploitation and selection, processing, thermal power generation and so on. Its holding parent company, Henan Shenhuo Group Co., Ltd., ranked 24th among the 2008 top 100 enterprises in China's coal industry.

Equity structure

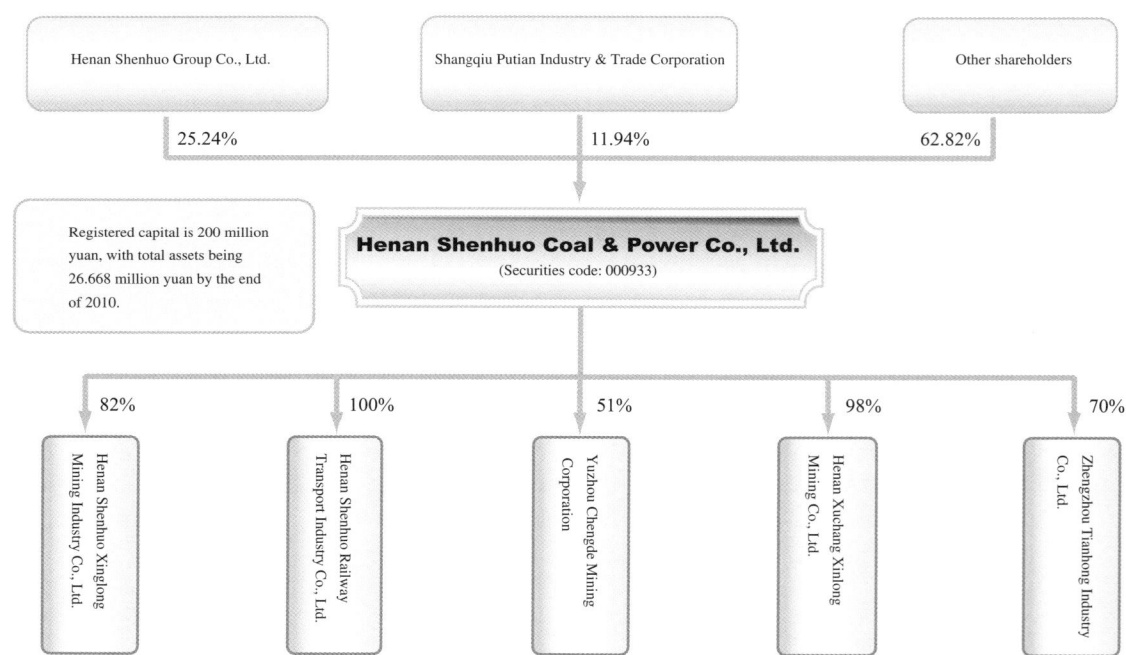

Business performance

2009

The company achieved a year-round operating revenue of 10.76 billion yuan, falling by 10.4% over the previous year in the same period. The net profit attributable to listed company shareholders was 600 million yuan, down 46% year on year.

The year-round raw coal production hit 5.72 million tons, up 19% year on year.

2010

The company's business performance showed a large-margin growth year on year: its year-round operating revenue totaled 16.90 billion yuan, up 57.1% year on year; the total profit hit 1.52 billion yuan, up 69.8% year on year, with 1.16 billion yuan of the net profit belonging to the listed company shareholders, up 92.9% over the previous year in the same period.

Within the year, it produced 7.112 million tons of coal, up 24.0% year on year; the average selling price of coal hit 730 yuan/ton, up 27.0% year on year.

Shanxi Xishan Coal & Electricity Power Co., Ltd.

Company overview

Xishan Coal & Electricity Power Co., Ltd., situated in Shanxi Province, China's large coal-resource province, is a listed coal-exploitation company boasting a relatively large scale in total share capital and assets. As the largest domestic coking coal producer, it mainly deals with coal production, exploitation and processing, power generation and marketing. The coal mines owned by the company boast a strong edge in resource reserves and scale.

Equity structure

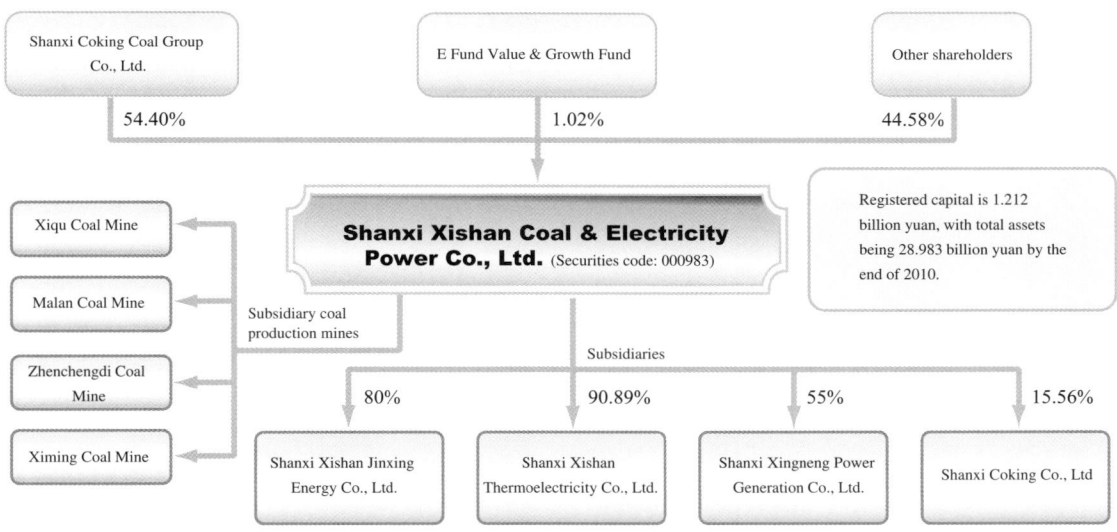

Business performance

2009

The company achieved a year-round operating revenue of 12.34 billion yuan, down 6.9% over the previous year; the net profit attributable to the parent company owners was 2.23 billion yuan, down 24.6% year on year. The decline in business performance was mainly due to a relatively substantial fall in the comprehensive selling price of coal.

The whole-year production and sales witnessed a soar: the raw coal output hit 18.556 million tons, up 14.6% year on year; the sales volume hit 17.796 million tons, up 10.0% year on year. The outputs of coking coal and fat clean coal were 5.592 million tons and 5.816 million tons respectively, up 1.9% and 7.2% year on year.

2010

The company achieved a year-round operating revenue of 16.62 billion yuan, up 36.7% year on year: coal sales revenue hit 14.99 billion yuan, up 38.4% year on year; sales revenue of electricity and thermal power totaled 1.19 billion yuan; that of coking coal was 6.52 billion yuan; the net profit attributable to the parent company hit 2.64 billion yuan, up 18.6% year on year.

In 2010, the company's raw coal production was 25.780 million tons, rising by 7.224 million tons year on year, substantially rising by 38.9%. This growth was mainly contributed by Xingxian Coal Mine.

The comprehensive selling price of coal in this year was 729.7 yuan/ton, up 19.9% year on year, the major driving force for the significant growth rate of its business performance, i.e. 36.7%.

Hebei Jinniu Energy Resources Co., Ltd.

Company overview

Hebei Jinniu Energy Resources Co., Ltd., located in Xingtai City, Hebei Province, listed on the Shenzhen Stock Exchange on September 9, 1999, takes quality 1/3 coking coal as its major type of coal, and mainly deals with coal exploitation, washing and selection, and coal marketing.

Equity structure

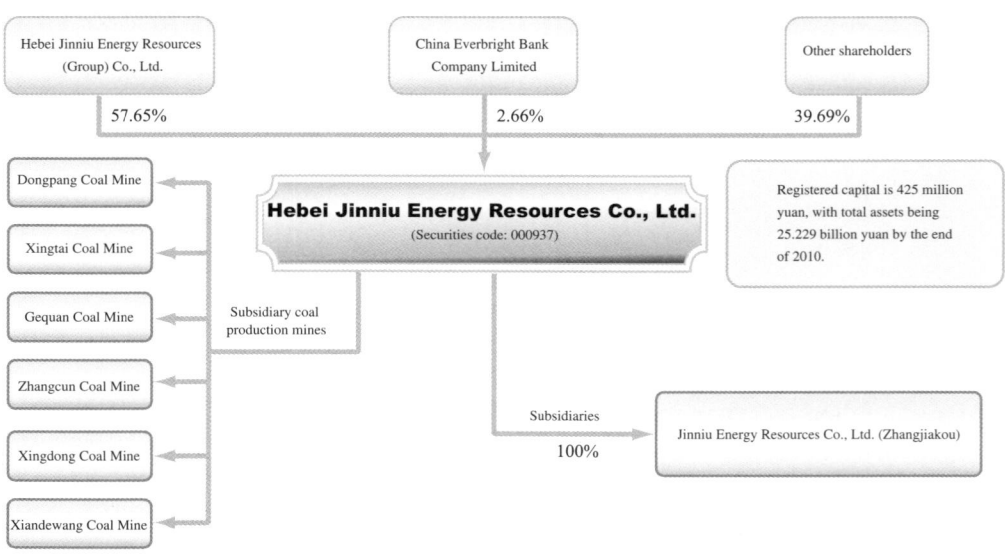

Business performance

2009

The company achieved a year-round operating revenue of 20.25 billion yuan, down 4.2% year on year, realizing a total operating profit of 2.25 billion yuan, down 39.2% year on year. The net profit attributable to the listed company shareholders was 1.61 billion yuan, down 39.3% year on year.

The all-year coal production was 25.8954 million tons, with washed clean coal making up 13.0775 million tons, a slight year-on-year growth.

In this year, the company made non-public additional acquisition of the mineral resources and mining equipment parts of Fengfeng Group, Handan Mining Group and Zhangjiakou Mining Group, all affiliated to Jizhong Energy Resources Co., Ltd., hence accomplishing the injection of group assets.

2010

The company brought off well the restructuring of major assets, achieving balanced development of four coal mines, that is, Xingtai, Fengfeng, Handan and Zhangjiakou, leading various indicators to hit new record highs.

The business performance saw a substantial growth, achieving a year-round operating revenue of 30.29 billion yuan, up 49.6% year on year. The total profit hit 3.32 billion yuan, up 47.6% year on year, realizing 2.39 billion yuan in net profit attributable to the listed company shareholders, up 43.2% year on year.

Raw coal production for the whole year totaled 31.02 million tons, with a year-on-year growth rate of up to 19.8%.

Shanghai Datun Energy Eesources Co., Ltd.

Company overview

Shanghai Datun Energy Resources Co., Ltd., registered in Pudong New Area, Shanghai, locates its major coal mines all in Xuzhou City, Jiangsu Province. It mainly deals with the domestic sale and export of prepared raw coal, washed clean coal, washed lump coal and the like, as well as self-run railway transport. Its holding parent company, China National Coal Group Corporation, ranked second among the 2008 top 100 enterprises in China's coal industry.

Equity structure

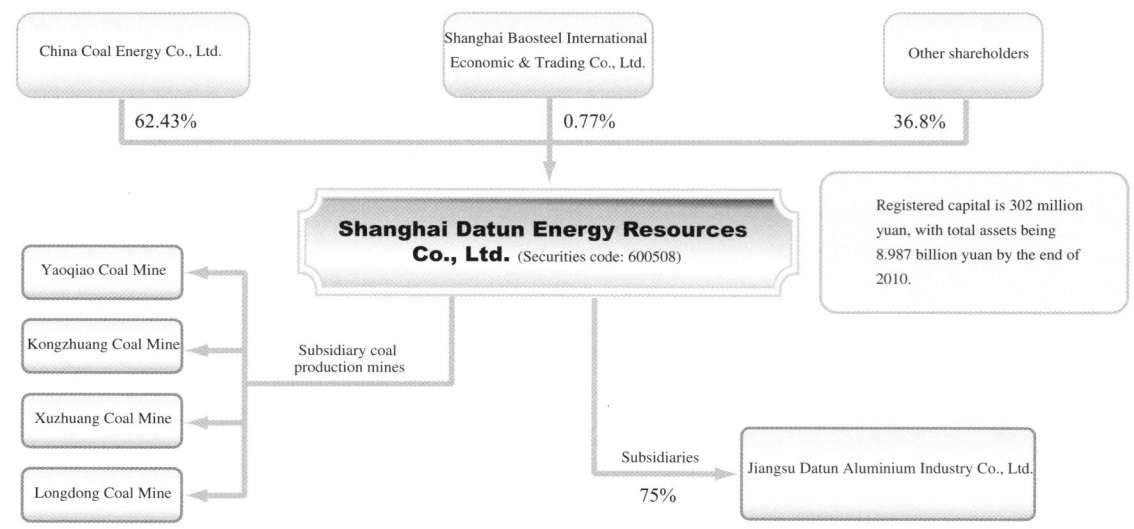

Business performance

2009

The company achieved a year-round operating revenue of 7.33 billion yuan, down 4.9% year on year, a total profit of 1.26 billion yuan, down 3.1% year on year, and a net profit of 0.97 billion yuan, down 4.1% year on year.

The year-round raw coal production totaled 8.609 million tons; that of washed clean coal was 2.780 million tons; electricity generation volume hit 2.77 kWh; and that of electrolytic aluminium was 0.107 million tons. Outputs of the major products all registered record highs.

2010

The company achieved a year-round operating revenue of 8.86 billion yuan, up 20.8% year on year, and a total profit of 1.74 billion yuan, up 37.7% year on year; meanwhile the net profit attributable to the parent company was 1.33 billion yuan, up 40.3% year on year.

Coal production hit a new high: in the whole year, raw coal production and clean coal production totaled respectively 9.092 million tons and 3.449 million tons, up 5.6% and 24.1% year on year. The electricity generation volume was 2.76 kWh; and the output of electrolytic aluminium was 0.109 million tons, up 1.6% year on year.

Kailuan Clean Coal Co., Ltd.

Company overview

Kailuan Clean Coal Co., Ltd., located in Tangshan City, Hebei Province, mainly deals with coal exploitation, raw coal washing and processing, and marketing of washed products. It was listed on June 2, 2004, on the Shanghai Stock Exchange.

Equity structure

Business performance

2009

The company achieved a year-round revenue of 10.977 billion yuan, up 16.8% year on year, and a net profit of 8230 million yuan, down 0.6% year on year.

　　Within the whole year, coal production was 7.7122 million tons, up 2.83% year on year, and that of clean coal hit 2.9196 million tons, up 17.15% year on year.

　　As for the coal chemical industry, coke production was 3.9514 million tons, up 29.52% year on year, and coal sales hit 3.9367 million tons, up 29.80% year on year.

2010

The company achieved a year-round revenue of 15.15 billion yuan, rising by 38.06% over the previous year, which was mainly due to the increase in the outputs and prices of some coal chemical products. The total profit was 1.14 billion yuan, falling by 2.6% over the previous year. The net profit attributable to the parent company shareholders was 0.87 billion yuan, up 5.6% over previous year.

As for the coal business, its production remained steady: the year-round raw coal production was 8.293 million tons, up 7.5% year on year; that of clean coal was 3.283 million tons, up 12.46% year on year.

The coal chemical business continued to expand: the year-round coke production hit 5.204 million tons, and the coke sales volume was 5.162 million tons, up 31.7% and 31.1% respectively over the previous year.

Taiyuan Coal Gasification Co., Ltd.

Company overview

Taiyuan Coal Gasification Co., Ltd., located in Taiyuan City, Shanxi Province, mainly deals with the production and marketing of coke, washed clean coal, raw coal, coal gas and coal chemical products. It was listed on June 22, 2000, on the Shenzhen Stock Exchange.

Equity structure

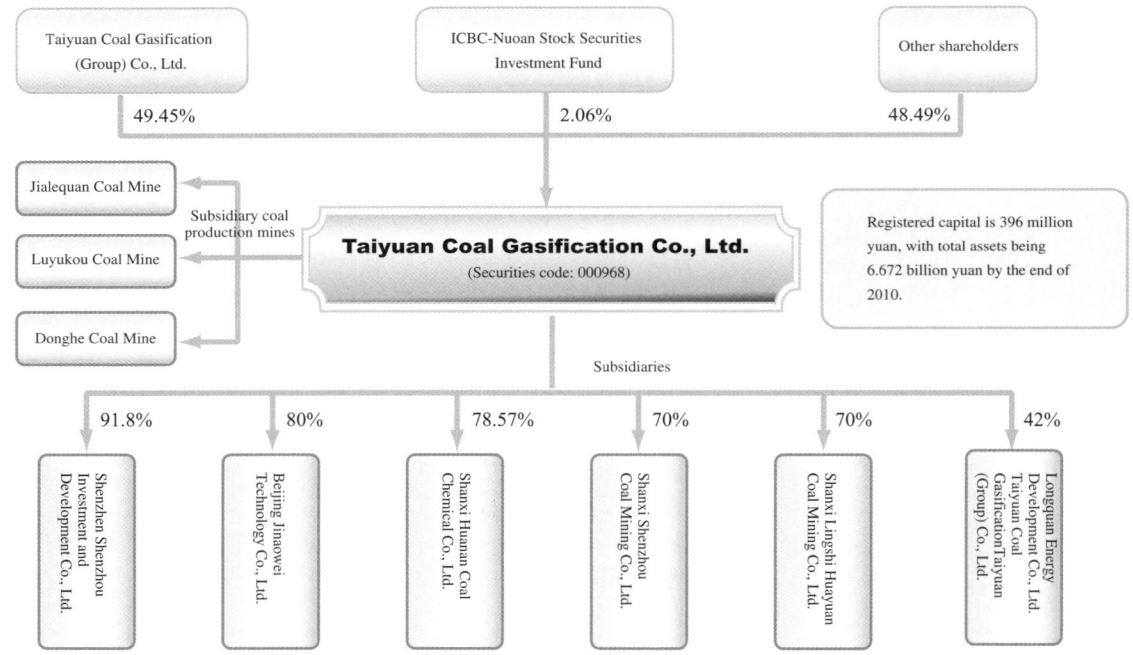

Business performance

2009

The company achieved a year-round revenue of 3.25 billion yuan, down 31.4% year on year, realizing a net profit of 0.38 billion yuan attributable to the parent company owners, which fell by 44.5% year on year. Since the share of coke in the major businesses approximated to 60%, the decline in coke prices was the primary factor in the drop in its business performance in this year.

2010

The company achieved a year-round operating revenue of 3.52 billion yuan, up 8.5% year on year; the operating profit was 0.47 billion yuan, down 15.9% year on year, with the net profit attributable to the parent company owners being 0.24 billion yuan, down 38.1% year on year. The recession of the market at the terminal of the industry chain was the major factor in the substantial decline in its business performance.

Chapter 10   Power Industry

Section 1: China's power construction and pattern

China has formed six large cross-regional power grids, i.e. North China, Northwest China, Northeast China, Central China, East China and South China, all of which have been interconnected.

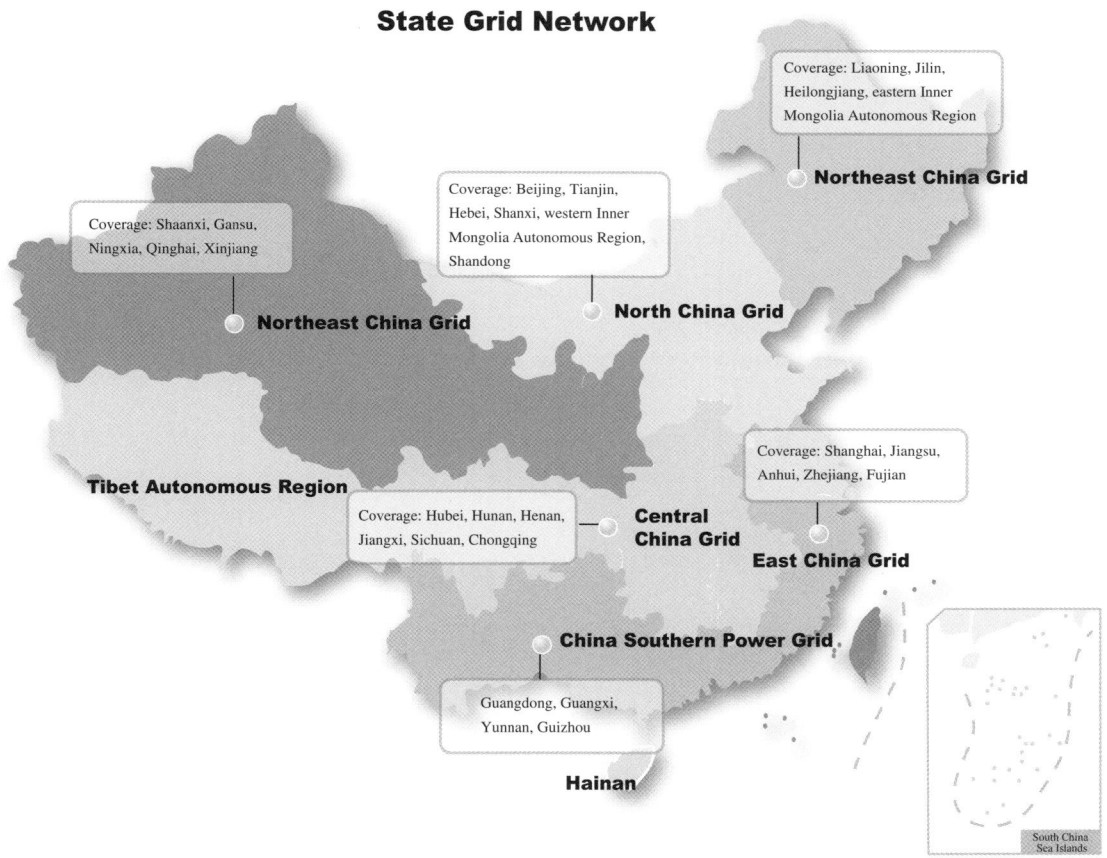

State Grid Network

Coverage: Liaoning, Jilin, Heilongjiang, eastern Inner Mongolia Autonomous Region

Northeast China Grid

Coverage: Shaanxi, Gansu, Ningxia, Qinghai, Xinjiang

Coverage: Beijing, Tianjin, Hebei, Shanxi, western Inner Mongolia Autonomous Region, Shandong

Northeast China Grid

North China Grid

Tibet Autonomous Region

Coverage: Shanghai, Jiangsu, Anhui, Zhejiang, Fujian

Coverage: Hubei, Hunan, Henan, Jiangxi, Sichuan, Chongqing

Central China Grid

East China Grid

China Southern Power Grid

Guangdong, Guangxi, Yunnan, Guizhou

Hainan

South China Sea Islands

In 2009, the completed investment in national power construction realized a relatively significant growth, with total investment in power infrastructure amounting to 755.84 billion yuan, up 19.9% year on year, namely, up 8.92 percentage points. Among this, grid construction witnessed a substantial increase, its completed investment being 384.71 billion yuan, making up 50.9% of total power construction investment, up 32.9% year on year. In contrast, the growth margin of power supply was slightly lower, with completed investment being 371.13 billion yuan, up 8.9% year on year, which was lower than the growth rate of nationwide power construction investment within the same period.

In 2010, national power construction investment completed was 705.10 billion yuan, down 8.5% over the previous year. Among this, the completed investment in power source construction was 364.10 billion yuan, down 4.3% over the previous year, and that of power grid investment was 341.00 billion yuan, down 12.5% over the previous year.

China's investment in power infrastructure

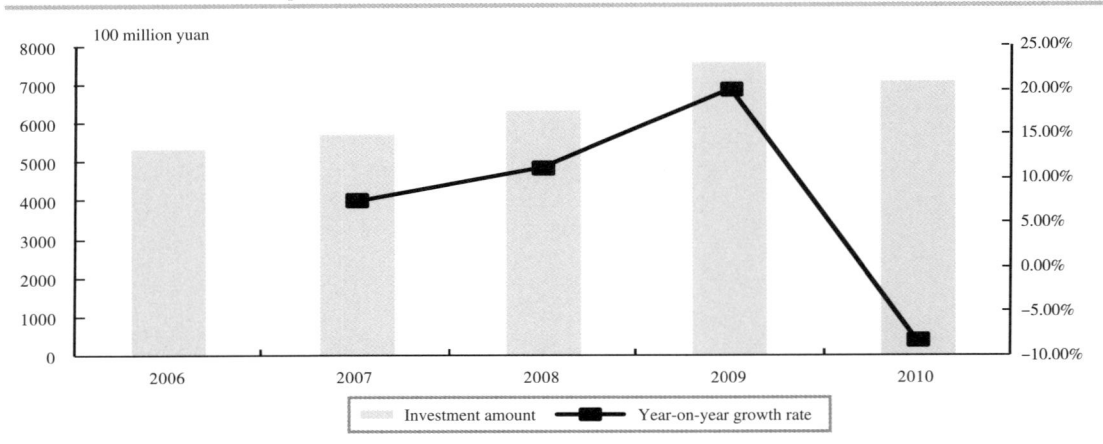

Data source: China power website, available at www.chinapower.com.cn

Composition of China's investment in power infrastructure

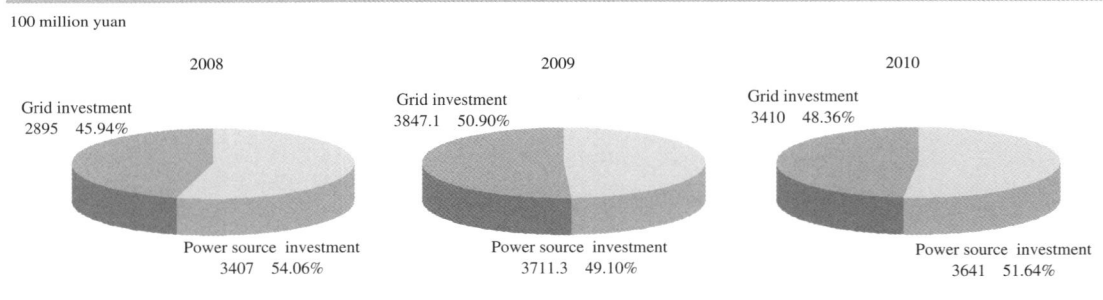

Data source: China power website, available at www.chinapower.com.cn

In the Eleventh Five-Year period, investment in power source infrastructure was inclined towards the field of power generation via non-fossil energy, and its structure represented a trend of accelerated adjustment. In 2010, among the completed investment in power source infrastructure, the share of hydropower, nuclear power and wind power each saw a salient increase, while that of thermal power continued to decrease.

Section 2: China's power supply and consumption

The production capability of China's power industry has been further improved. By the end of 2010, the capacity of power generation equipment nationwide was 962.19 million kW,

Composition of investment in China's power construction during the years 2008–2010

100 million yuan

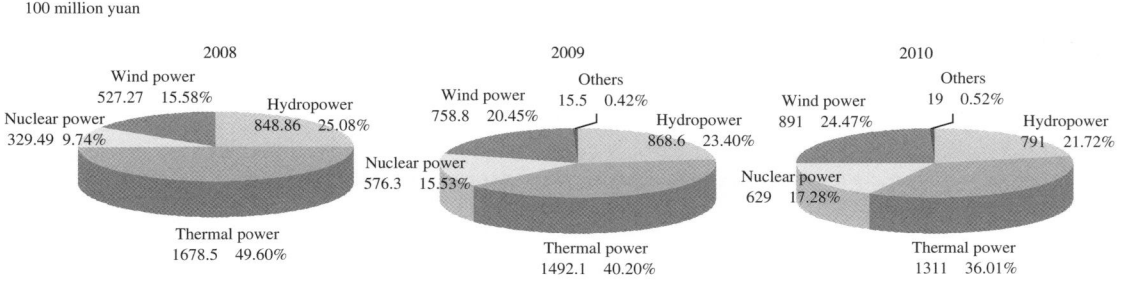

Data source: China power website, available at www.chinapower.com.cn

among which hydropower installed capacity was 213.40 million kW, that of thermal power was 706.63 million kW, that of nuclear power installed capacity was 10.82 million kW, and that of wind power grid integration was 31.07 million kW. During the Eleventh Five-Year period, the share of thermal power installed capacity experienced a decline year by year, while the installation share of non-fossil energy such as hydropower, nuclear power and wind power boasted a rise of up to 26.5%.

China's power equipment capacity

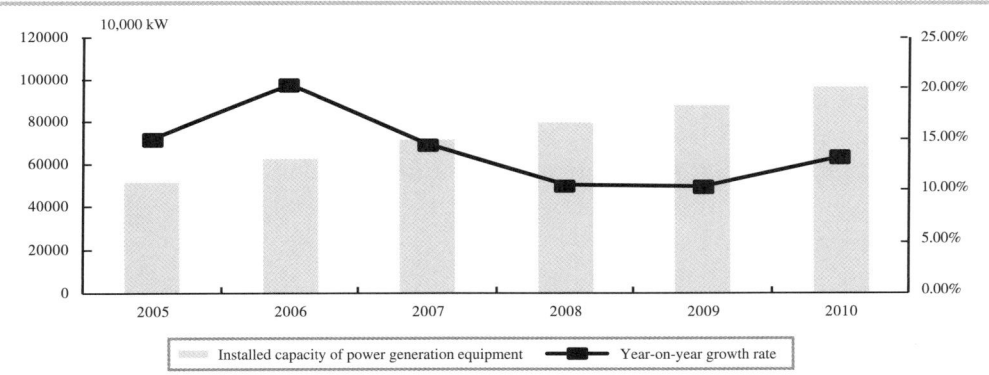

Data source: China power website, available at www.chinapower.com.cn

Composition of China's power equipment capacity during the years 2006–2010

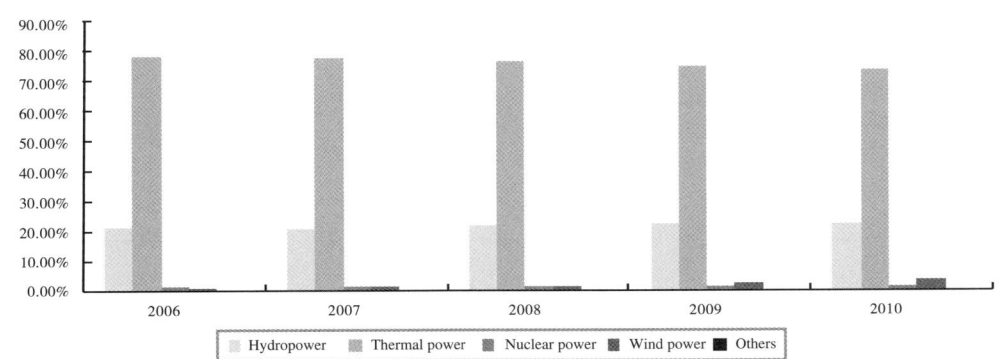

Data source: China power website, available at www.chinapower.com.cn

In terms of the structure of power supply, the installed capacity of power generation nationwide, the growth rate of thermal power declined relatively rapidly, while hydropower maintained a relatively fast growth and wind power installation saw a high-speed growth. In 2010, hydropower, thermal power and wind power all witnessed a year-on-year increase, by 8.7%, 8.6% and 77.1% respectively.

China's power equipment capacity during the years 2006–2010

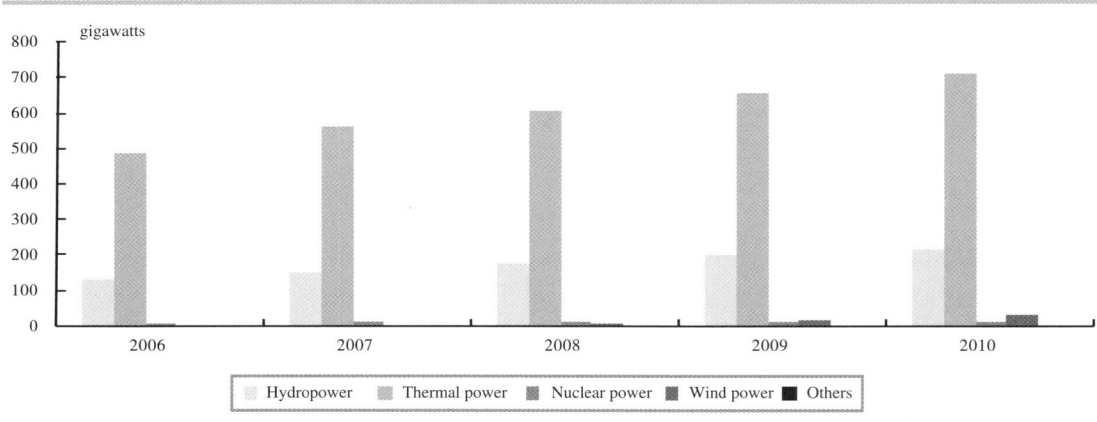

Data source: China power website, available at www.chinapower.com.cn

In 2010, the newly added production capability of infrastructure nationwide continued to maintain a large scale, with the capacity of newly added power generation equipment in infrastructure hitting 91.27 million kW. Among this newly added power equipment capacity, the shares of hydropower and thermal power underwent a fall, by 4.0 and 3.4 percentage points respectively, and that of wind power witnessed a rise of 5.3 percentage points.

Newly added capacity of power generation equipment in China's infrastructure

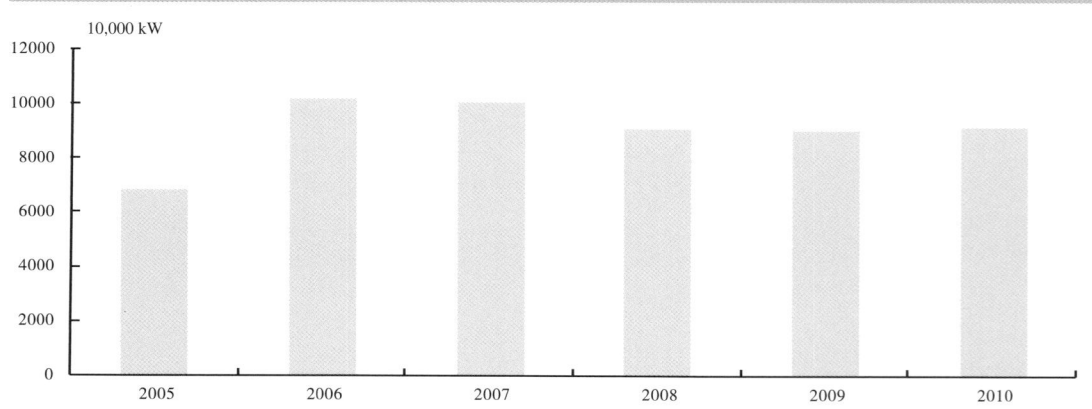

Data source: China power website, available at www.chinapower.com.cn

Composition of newly added installed capacity of power generation in China's infrastructure in 2009

10,000 kW

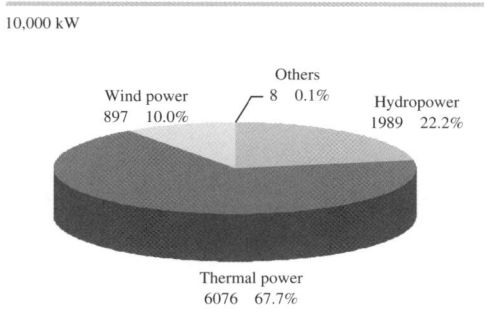

Composition of newly added installed capacity of power generation in China's infrastructure in 2010

10,000 kW

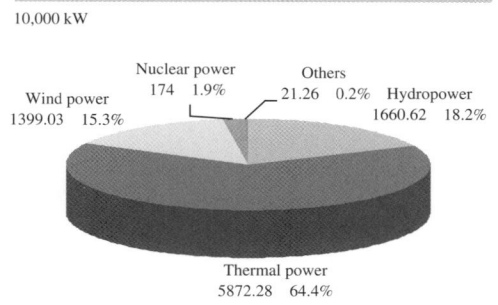

Data source: China power website, available at www.chinapower.com.cn

Power generation witnessed an accelerated rebound. In 2009, the nationwide full-scale electricity generation amounted to 3650.6 billion kWh, up 7.0% year on year. Among this, that of thermal power made up 2981.4 billion kWh, up 7.2%, and that of hydropower 55.45 billion kWh, up 4.3% year on year. In 2010, the electricity generation nationwide hit 4141.3 billion kWh, up 13.4% year on year, its growth rate rising by 6.4 percentage points over the previous year. In the Eleventh Five-Year period, the shares of clean-energy sources in electricity generation such as hydropower, nuclear power and wind power witnessed a significant increase.

Full-scale electricity generation

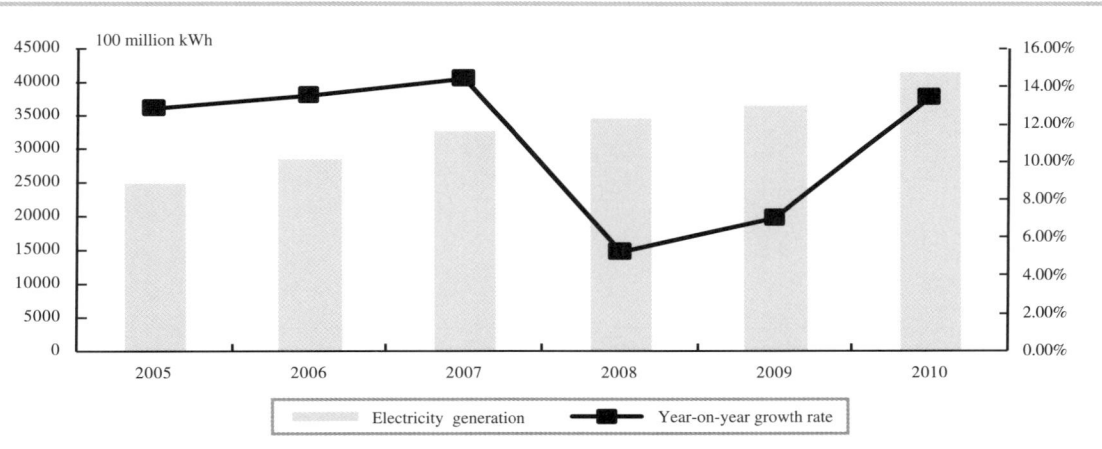

Data source: China power website, available at www.chinapower.com.cn

Composition of China's electricity generation in 2009

100 million kWh

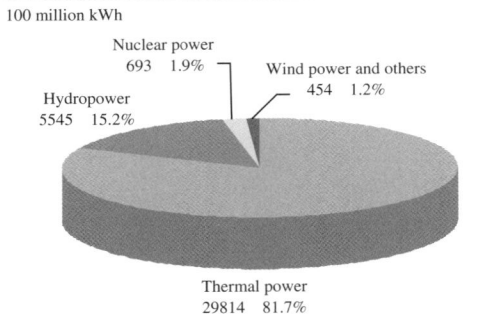

Nuclear power
693   1.9%

Wind power and others
454   1.2%

Hydropower
5545   15.2%

Thermal power
29814   81.7%

Composition of China's electricity generation in 2010

100 million kWh

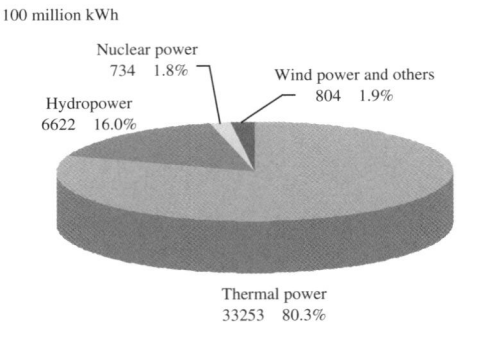

Nuclear power
734   1.8%

Wind power and others
804   1.9%

Hydropower
6622   16.0%

Thermal power
33253   80.3%

Data source: National Energy Administration

Looking at the utilization efficiency of power generation equipment, under the effect of the slowdown in the growth rate of power utilization and the operation of a great mass of newly added power supply projects, the average utilization hours of power generation equipment saw a fall in recent years. In 2009, the cumulative utilization hours of power generation equipment were 4527 h, down 121 h year on year; while in 2010 this cumulative hourage hit 4660 h, up 133 h over the previous year and the first rebound after a continuous decline since 2004.

Utilization hours of power generation equipment in power plants over 6000 kWh

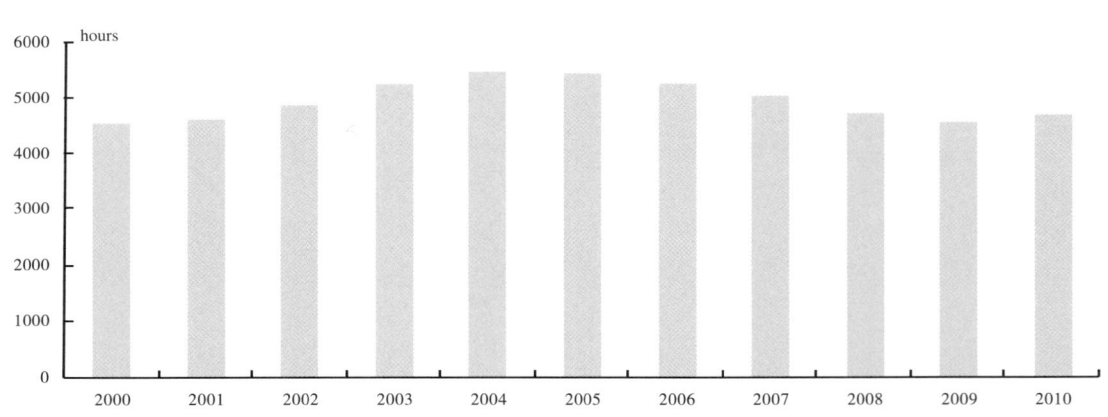

Data source: China power website, available at www.chinapower.com.cn

Looking at the structure of power supply, the utilization efficiency of power generation equipment demonstrated features as follows: during the years 2005–2008, utilization hours of thermal power fell at a relatively rapid rate, the declining rate of which slightly reduced in 2009, and stated to recover in 2010; as for hydropower, the utilization hours saw a year-on-year growth both in 2007 and 2008, but then in 2009 experienced an obvious year-on-year decline, which again rebounded in 2010; in contrast, in the case of nuclear power, the utilization hours remained generally steady.

Composition of utilization hours of power generation equipment in power plants above 6000 kWh

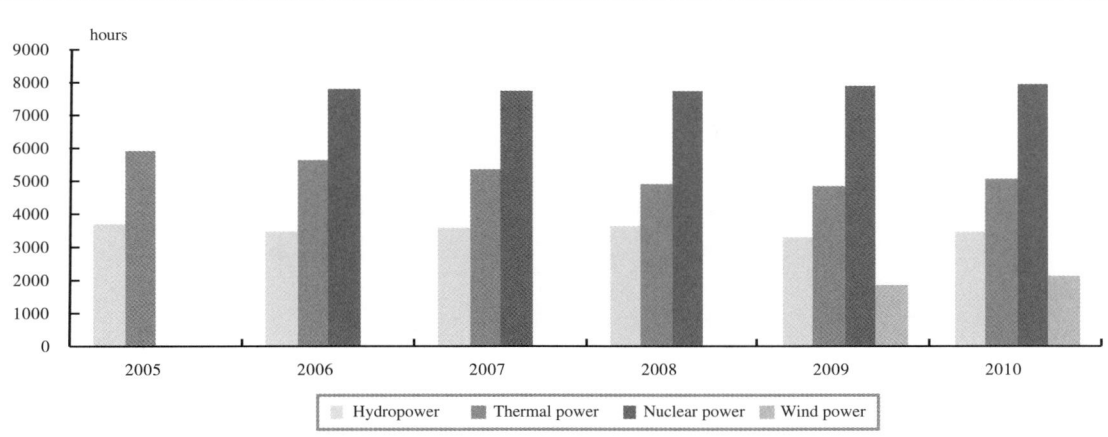

Data source: China power website, available at www.chinapower.com.cn

In 2009, as the macro economy recovered in a steady manner, domestic power demand kept increasing. Power consumption of the whole society the year round hit 3643 billion kWh, up 6.0% year on year, with the growth rate rising by 0.8 percentage points over the previous year. In 2010, the year-round power consumption the previous was 4192.3 billion kWh, up 15.1% over year, with the annual average of growth rate during the Eleventh Five-Year period being 11.09%.

Power consumption of the whole society in China

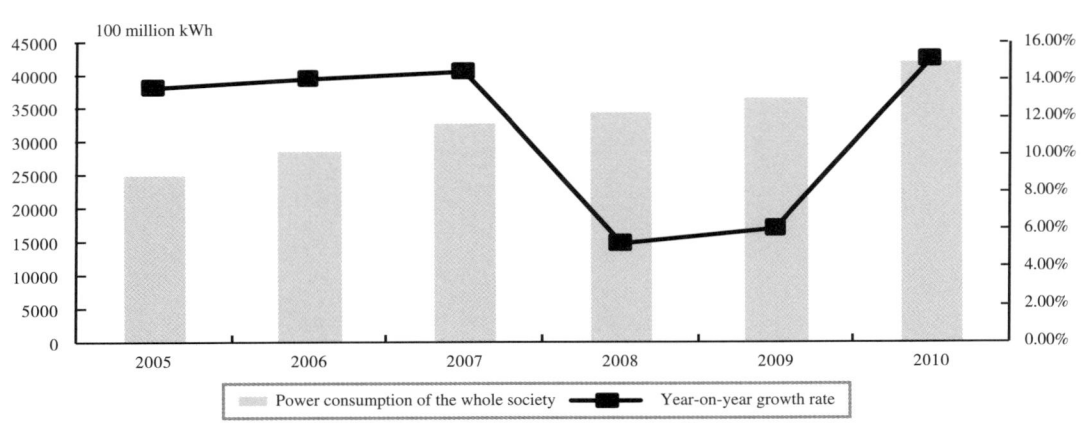

Data source: China power website, available at www.chinapower.com.cn

Looking at different industries, the fluctuation in growth rate was relatively insignificant in the case of the primary and tertiary industries and household electricity consumption, while power demand in the secondary industry was rather sensitive to the influence of the macro economy. In 2010, the power consumption of the primary industry was 98.4 billion kWh, at an annual average growth rate of 5.4% during the Eleventh Five-Year

Comparison of China's power consumption in different industries

Unit: 100 million kWh

	Primary industry	Year-on-year growth rate of primary industry power consumption	Secondary industry	Year-on-year growth rate of secondary industry power consumption	Tertiary industry	Year-on-year growth rate of tertiary industry power consumption	Household electricity consumption	Year-on-year growth rate of household electricity consumption
2005	741	7.6%	18478	13.4%	2631	12.9%	2938	16.2%
2006	832	9.9%	21354	14.3%	2822	11.8%	3240	14.7%
2007	860	5.2%	24847	15.7%	3167	12.1%	3584	10.6%
2008	879	1.9%	25863	3.8%	3498	9.7%	4035	11.8%
2009	947	7.9%	26993	4.2%	3921	12.1%	4571	11.9%
2010	984	3.9%	31318	16.0%	4497	14.7%	5125	12.1%

period; that of the secondary industry hit 3131.8 billion kWh, at an annual average growth rate of 10.9% during the Eleventh Five-Year period; and that of the tertiary industry and household electricity consumption were 449.7 billion kWh and 512.5 billion kWh respectively, at an annual average growth rate of 12.3% and 12.7% respectively during the Eleventh Five-Year period.

In terms of the year-round trend, in 2009, the power consumption of the whole society had the following characteristics: a sluggish start, steady improvement, recovery month by month and an accelerated growth. In the first half of the year, the impact exerted by the financial crisis still lingered; then from the second half of the year onwards, the macro economy kept steady and took a turn for the better, thus driving the increase in power consumption nationwide in that the year-on-year growth of the monthly power consumption of the whole society turned from negative into positive, with the growth rate continuing to increase. In 2010, influenced by the previous year's base and the strengthened execution of energy-saving and emission reduction policies, the power consumption of the whole society showed a high-speed growth during the first quarter, then was steady during the months April to August, and from then on started to fall back.

Monthly power consumption of the whole society in 2009

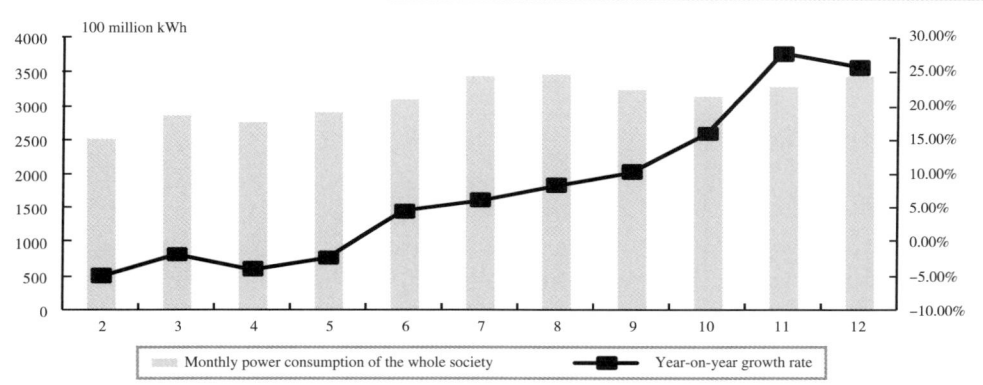

Data source: China power website, available at www.chinapower.com.cn

Monthly power consumption of the whole society in 2010

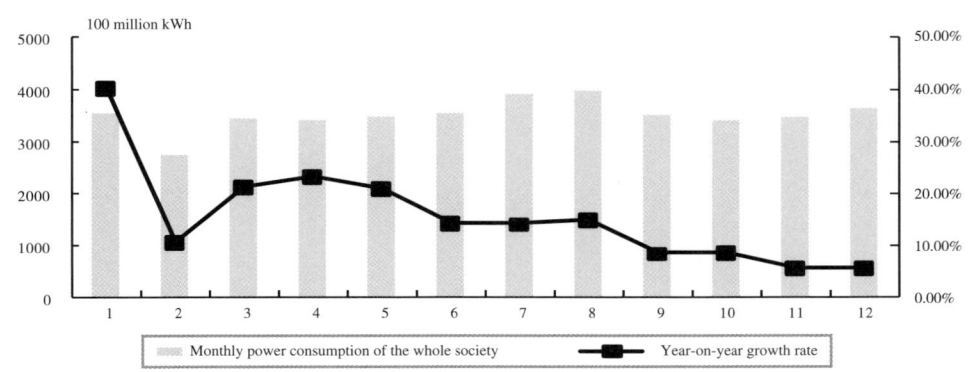

Data source: China power website, available at www.chinapower.com.cn

Section 3: Economic benefits and competition pattern of China's power industry

In 2010, during the first three quarters the booming index of the power industry continued to go up, but in the fourth quarter it began to descend. In the fourth quarter this booming index indicated 100.32 (100 in the year 2001), falling by 1.39 points compared with that in the third quarter, thus stopping the continuous increase for four quarters and starting to go down.

Booming index during the years 2002–2010

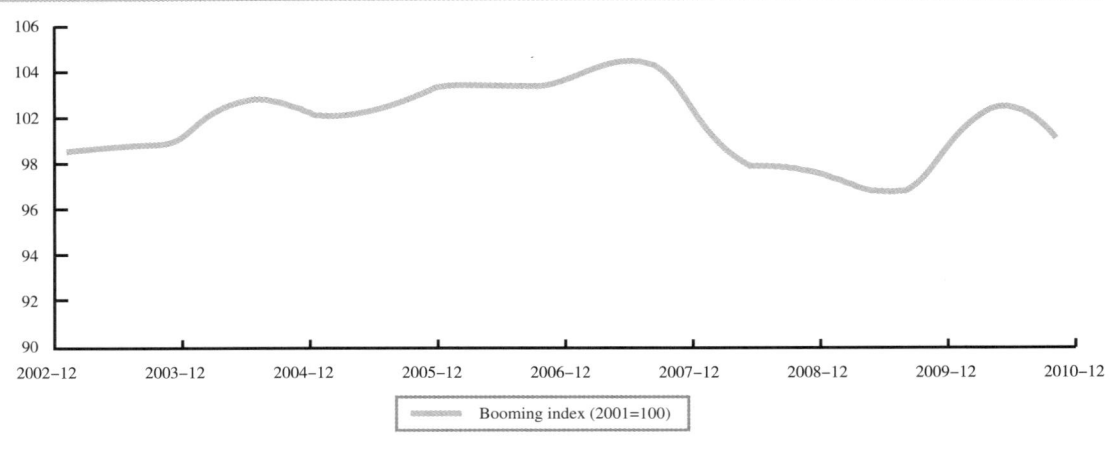

Data source: http://www.ce.cn/

During the fourth quarter of 2010, of the 10 indicators comprising the early warning index of the power industry, two indexes were located in the "red light range", four in the "green light range", one in the "pale blue light range" and one in the "dark blue range". In this period, the early warning index of the power industry read 30 points, continuing the descending trend for a second quarter, yet remaining stable, fluctuating within the "green light range".

Booming index lights

Index Name	2007	2008				2009				2010			
	12	3	6	9	12	3	6	9	12	3	6	9	12
Total-profit growth rate													
Total-tax growth rate													
Annual average rate of employee development													
Producer's price index													
Growth rate of crude oil production													
Inventory turnover ratio													
Receivables turnover ratio													
Rate of sales income growth													
Rate of fixed investment growth													
Exports growth rate													
Early warning index	27.0	22.0	23.0	26.0	21.0	17.0	19.0	19.0	24.0	31.0	34.0	33.0	30.0

Source: Center for China Economy Industry Boom Index Research, http://www.ce.cn/

In recent years the asset-liability ratio of power enterprises went up year by year, severely affecting the sustainability of the power industry.

Assets and liabilities in the power generation industry

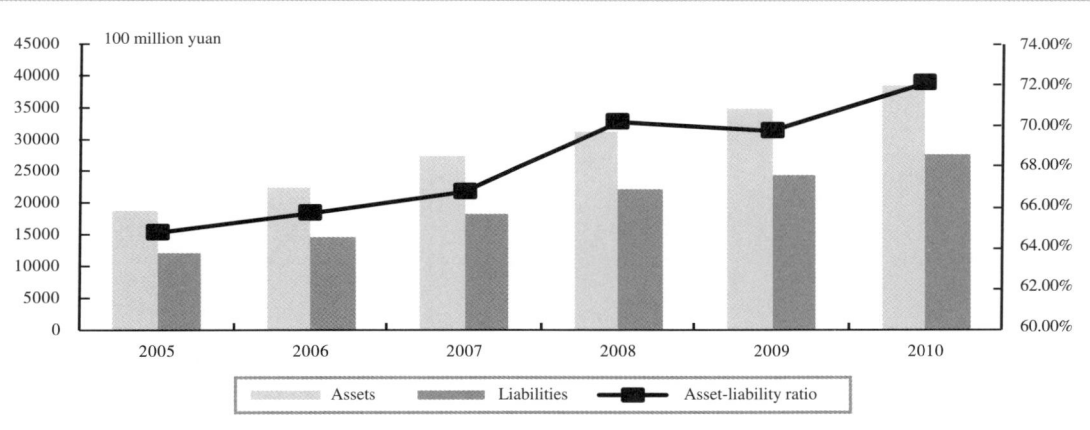

Data source: NBS

165

Assets and liabilities in the power supply industry

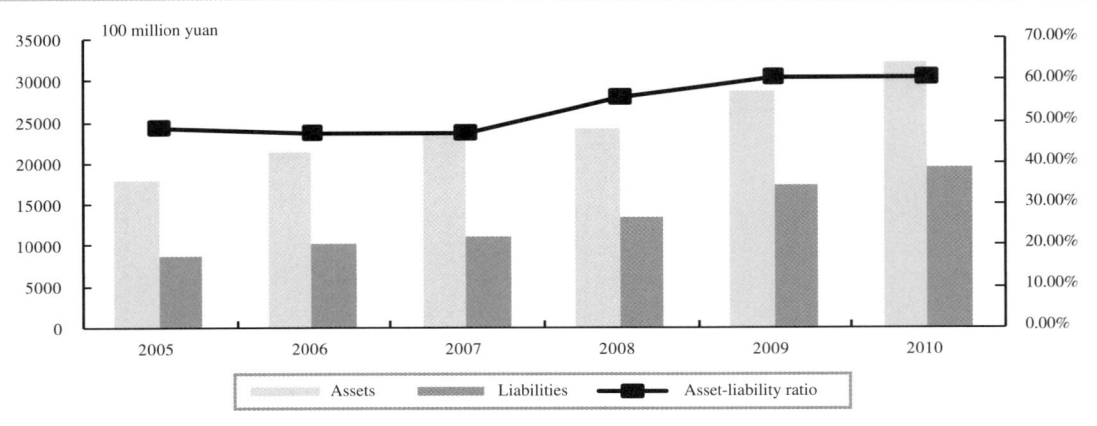

Data source: NBS

During the months January to November of 2009, the power generation industry achieved a profit of 82.79 billion yuan, rising drastically by 49.9 times, the main reasons being as follows: first, in 2008 the low temperature and natural disasters such as heavy rain and snow and freezing hazards baffled the relevant technology; second, the fallback of coal prices during the first half of 2009 reduced the production costs.

During the months January to November of 2010, the power generation industry achieved a profit of 82.69 billion yuan, but influenced by the base of the previous year, its growth rate witnessed a fallback, up 3.6% year on year.

Profits in the power generation industry

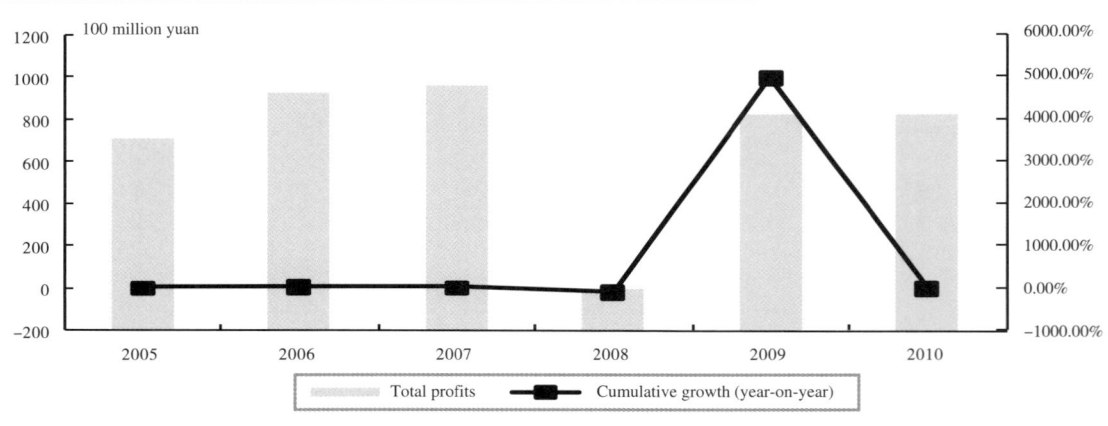

Data source: NBS

Within the months January to November of 2009, the cumulative profit of the power supply industry hit 6.30 billion yuan, realizing the turn from loss into gain, yet still falling by 77.8% compared with that of the previous year in the same period.

During the months January to November of 2010, the profitability of the power supply industry continued to get better, realizing a profit of 59.21 billion yuan, increasing greatly by 1828.1% year on year.

Profits in the power supply industry

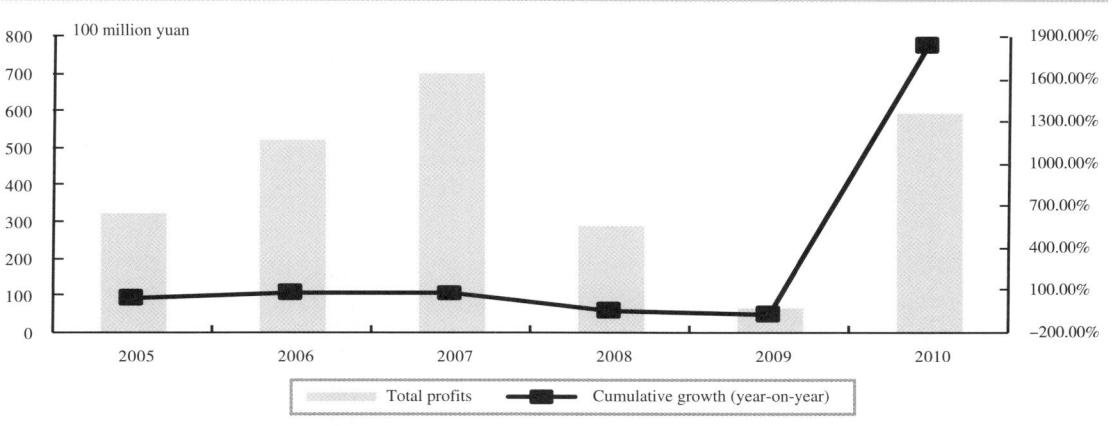

Data source: NBS

China's top five power provider groups

In 2009, China's top five power provider groups achieved a substantial growth in sales revenue, all surpassing the mark of 100 billion yuan. China Huaneng Group realized a sales revenue of 178.7 billion yuan, up 18% year on year; China Datang Group had a sales revenue of 151.6 billion yuan, up 49% year on year; for China Guodian (Group) Corporation (China GuoDian), its sales revenue hit 125.2 billion yuan, a record for the company; as for China Huadian Corporation (Huadian Group) and China Electricity Investment Co., Ltd. (CPI Group), their sales revenues both surpassed the 100-billion-yuan mark for the first time, hitting 103.0 billion yuan and 101.2 billion yuan respectively, both with substantial year-on-year growth.

In 2010, the sales revenues of these top five power provider groups continued to maintain a relatively significant growth: Huaneng Group's sales revenue surpassed the 200-billion-yuan mark; Datang Group and China GuoDian entered the rankings of Fortune Global 500, their sales revenues hitting 177.1 billion yuan and 165.4 billion yuan respectively; the annual sales revenue growth of Huadian Group and CPI Group both surpassed 20 billion yuan.

Sales revenues of top five power provider groups

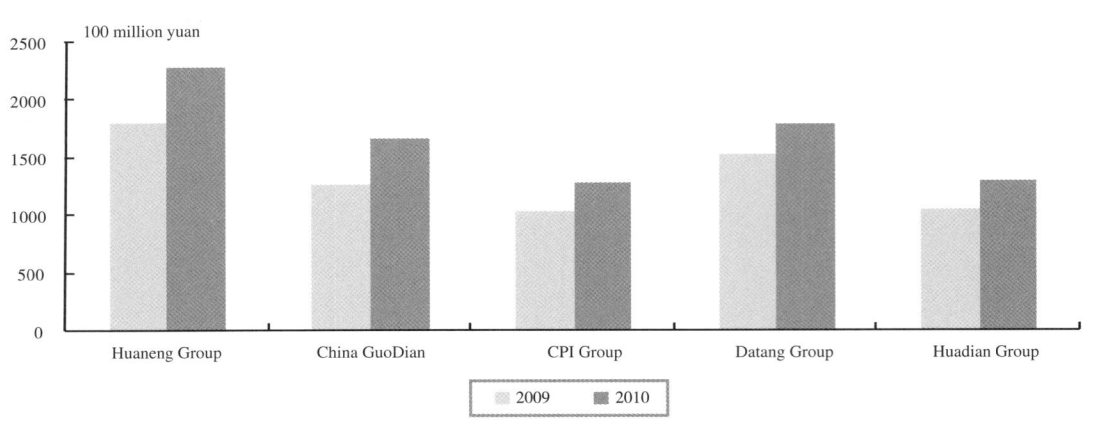

Data source: Company annals

In 2009, the top five power provider groups experienced a significant turn from loss to gain: their total profit for the whole year hit 19.601 billion yuan. Among this, Huaneng Group had 6.09 billion yuan, Huadian Group 2.09 billion yuan, Datang Group 2.28 billion yuan, China GuoDian 5.98 billion yuan and CPI Group 3.17 billion yuan.

In 2010, due to the influence of the increase in coal prices, the total profit of these five group corporations as a whole didn't witness a large-scale growth. Huaneng Group's total profits still ranked first among them; the total profit of CPI Group boasted the largest growth margin, which was also the highest since its founding.

Total profits of top five power provider groups

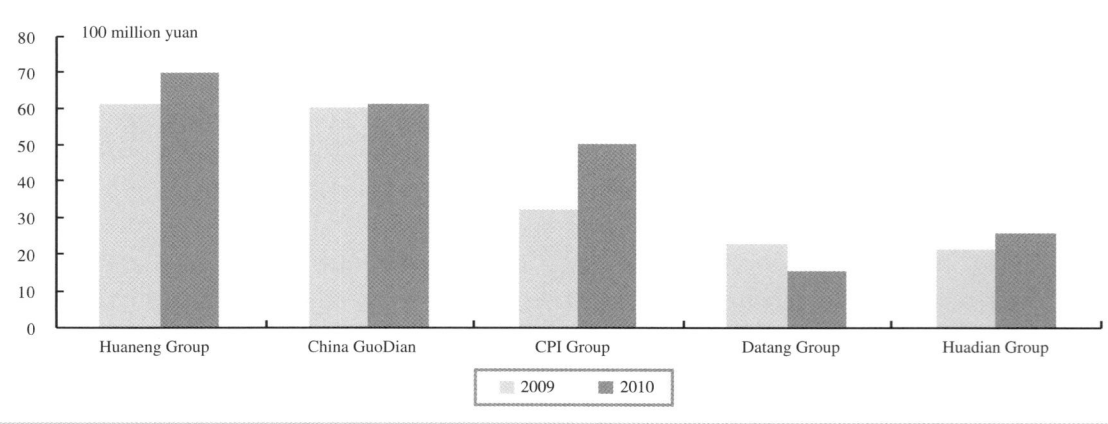

Data source: Company annals

In 2009, the installed capacity of the five leading power provider groups kept a relatively rapid growth, accounting for 48.1% of the total installed capacity nationwide. Specifically, Huaneng Group's and Datang Group's installed capacity both surpassed the 100-million kW mark.

In 2010, the installed capacity of the five groups continued to grow, among which that of Huaneng Group and Datang Group both surpassed 100 million kW.

Installed capacity of top five power provider groups

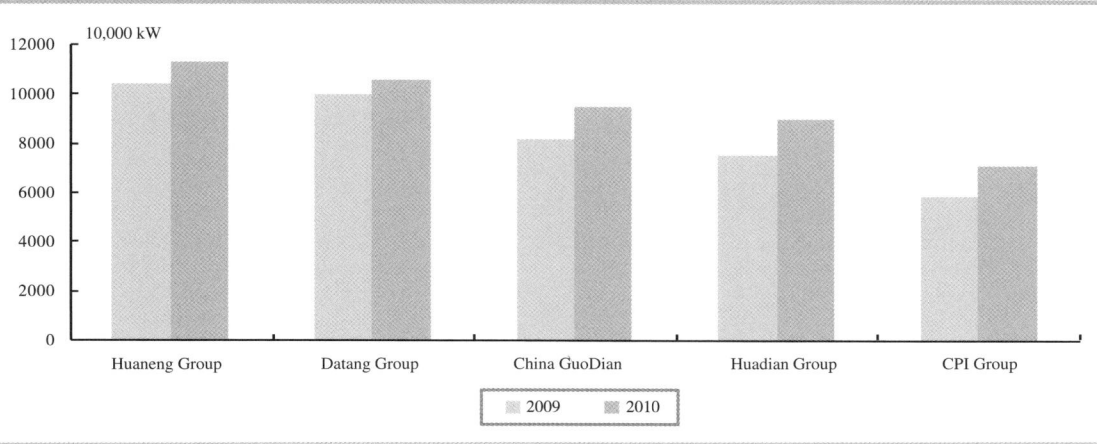

Data source: Company annals

In 2009, the electricity generation of these five power provider groups generally showed a trend of recovering after a weak start: during the first two quarters, it declined drastically, then steadily went up in the third quarter, and finally experienced a substantial growth in the fourth quarter. Specifically, CPI Group and China GuoDian boasted comparatively high growth rates of power generation.

In 2010, the electricity generation of the top five power provider groups continued the increasing trend, among which that of Huaneng Group was the most significant, with its electricity generation hitting 53.7 billion kW, and its growth rate up 28.0% year on year.

Electricity generation of top five power provider groups in 2009

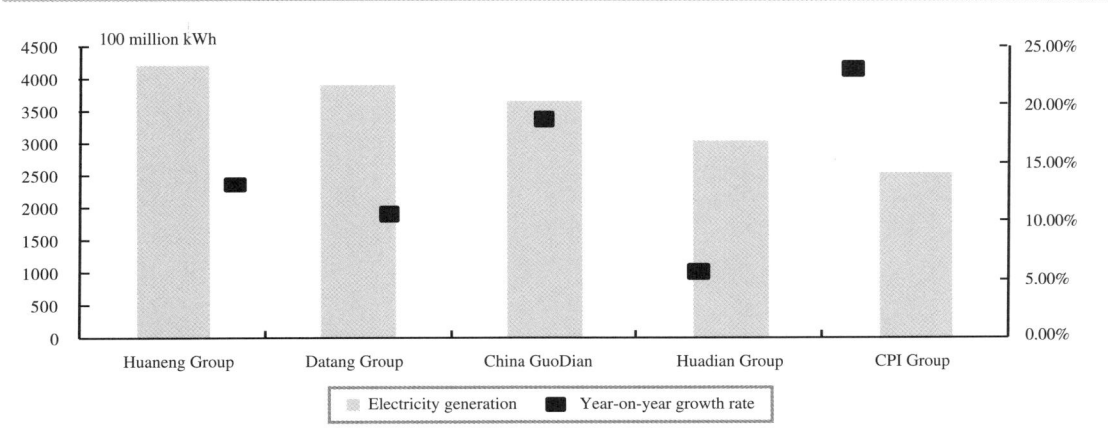

Data source: Company annals

Electricity generation of top five power provider groups in 2010

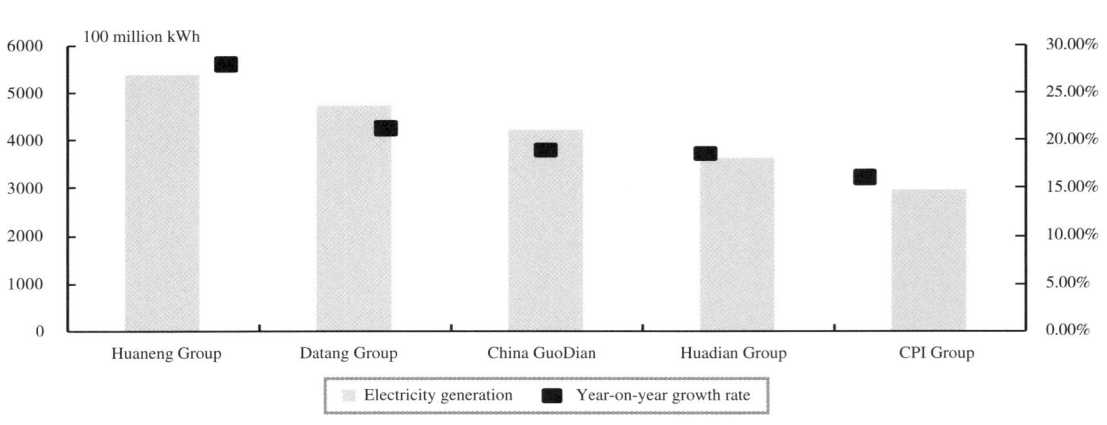

Data source: Company annals

Section 4: Development of power sub-industries in China

Thermal power industry

In 2009, China's installed capacity of thermal power continued to grow, which by the year end achieved a year-round total of 651.08 million kW, up 8.0% year on year, making up 74.6% of the total installed capacity nationwide. The all-year electricity generation via thermal power hit 2944.7 kWh, up 7.2% year on year, making up 81.7% of the total electricity generation and still keeping the leading place.

In 2010, the nationwide installed capacity of thermal power continued to maintain growth, which by the year end amounted to 706.63 million kW, up 8.5% year on year, making up 73.4% of the total installed capacity. The whole-year thermal power generation hit 3295.8 kWh, up 11.9% year on year, making up 80.3% of total electricity generation and remaining in first place.

Installed capacity of thermal power in China

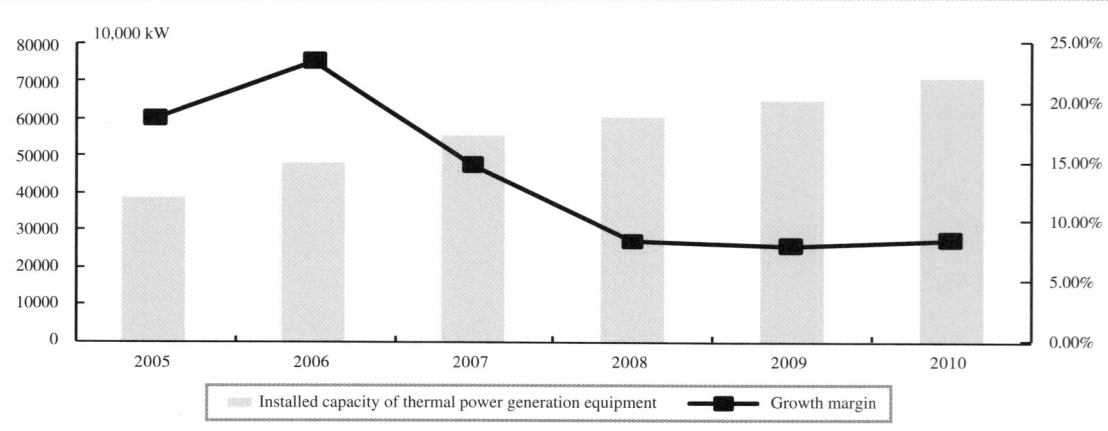

Data source: China Electricity Council

China's thermal power generation

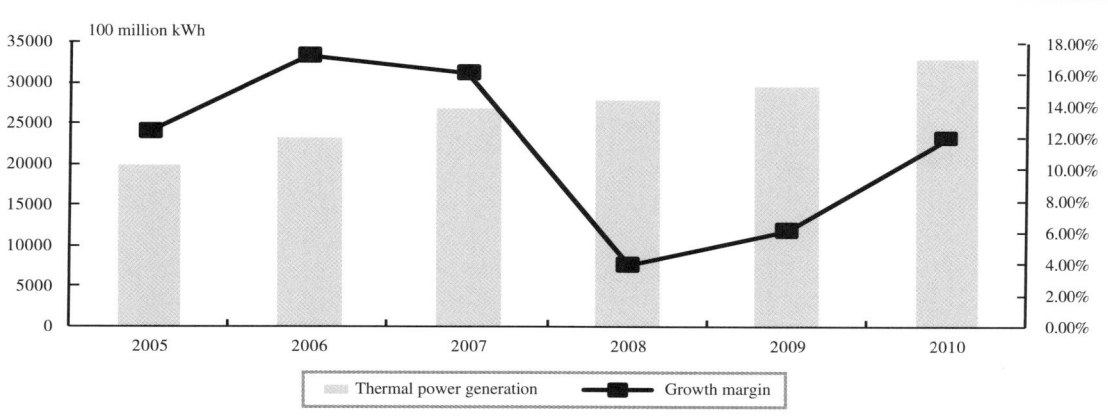

Data source: China Electricity Council

During the recovery from the financial crisis, the average utilization hours of the installed thermal power equipment in 2009 were 486.5 h, a record low in recent years; in 2010, the average utilization hours of the installed thermal power equipment witnessed a growth.

Average utilization hours of thermal power equipment

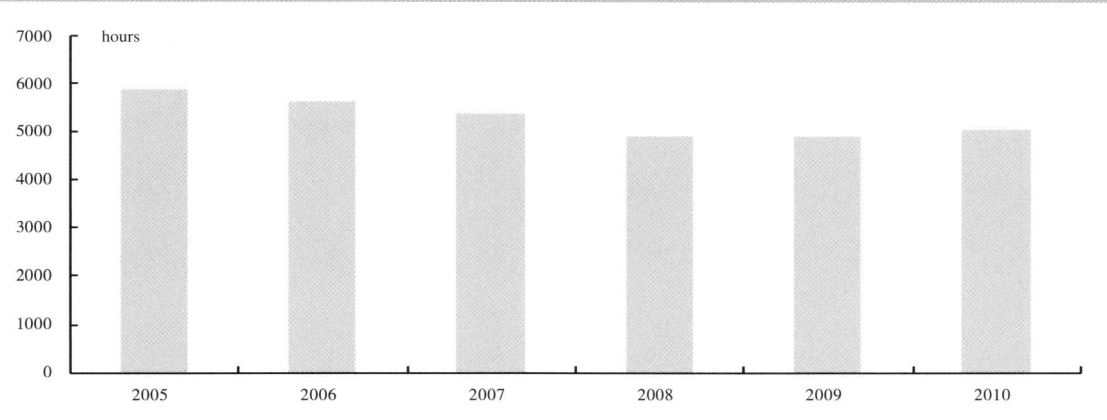

Data source: China Electricity Council

Hydropower industry

By the end of 2009, hydropower installed capacity hit 196.29 million kW, up 13.7% year on year, making up 22.5% of total installed capacity, with its growth rate higher than the average level. By the end of 2010, the hydropower installed capacity nationwide was 213.40 million kW, up 8.7% year on year, accounting for 22.2% of the total installed capacity.

Installed capacity of hydropower electricity generation

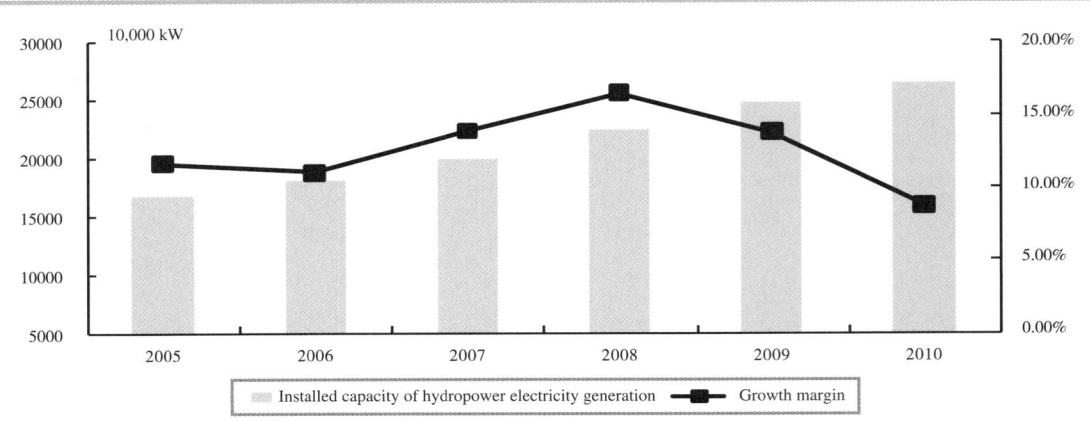

Data source: China Electricity Council

In 2009, the whole-year hydropower electricity generation was 553.5 billion kWh, up 5.9% year on year, making up 14.3% of the total electricity generation; in 2010, the year-round hydropower electricity generation hit 659.4 billion kWh, up 19.1% year on year, making up 16.0% of the total.

Hydropower electricity generation

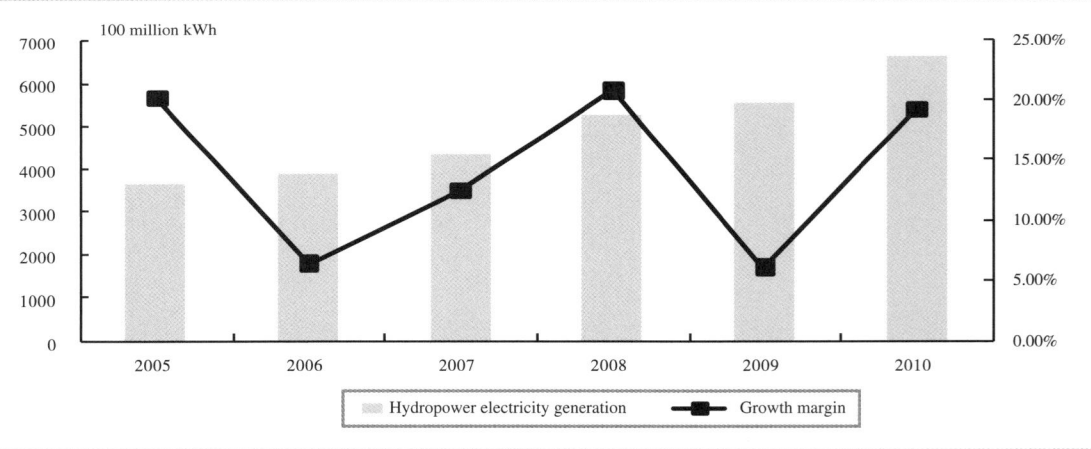

Data source: China Electricity Council

In 2009, the utilization hours of hydropower generation equipment were 3328 h, rising by 261 h over the previous year; in 2010 the utilization hours saw an increase, up to 3429 h.

Average utilization hours of hydropower generation equipment

hours
3700
3600
3500
3400
3300
3200
3100
2005
2006
2007
2008
2009
2010

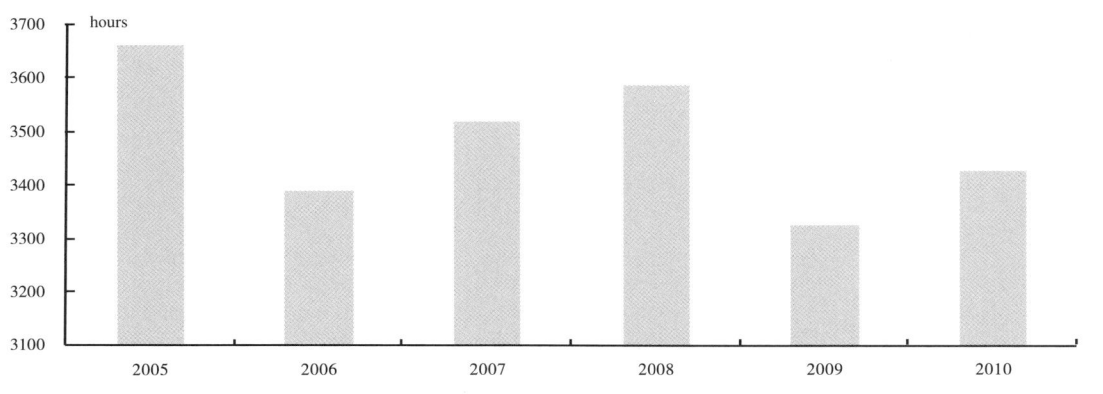

Data source: China Electricity Council

Plan of China's 13 largest hydropower bases

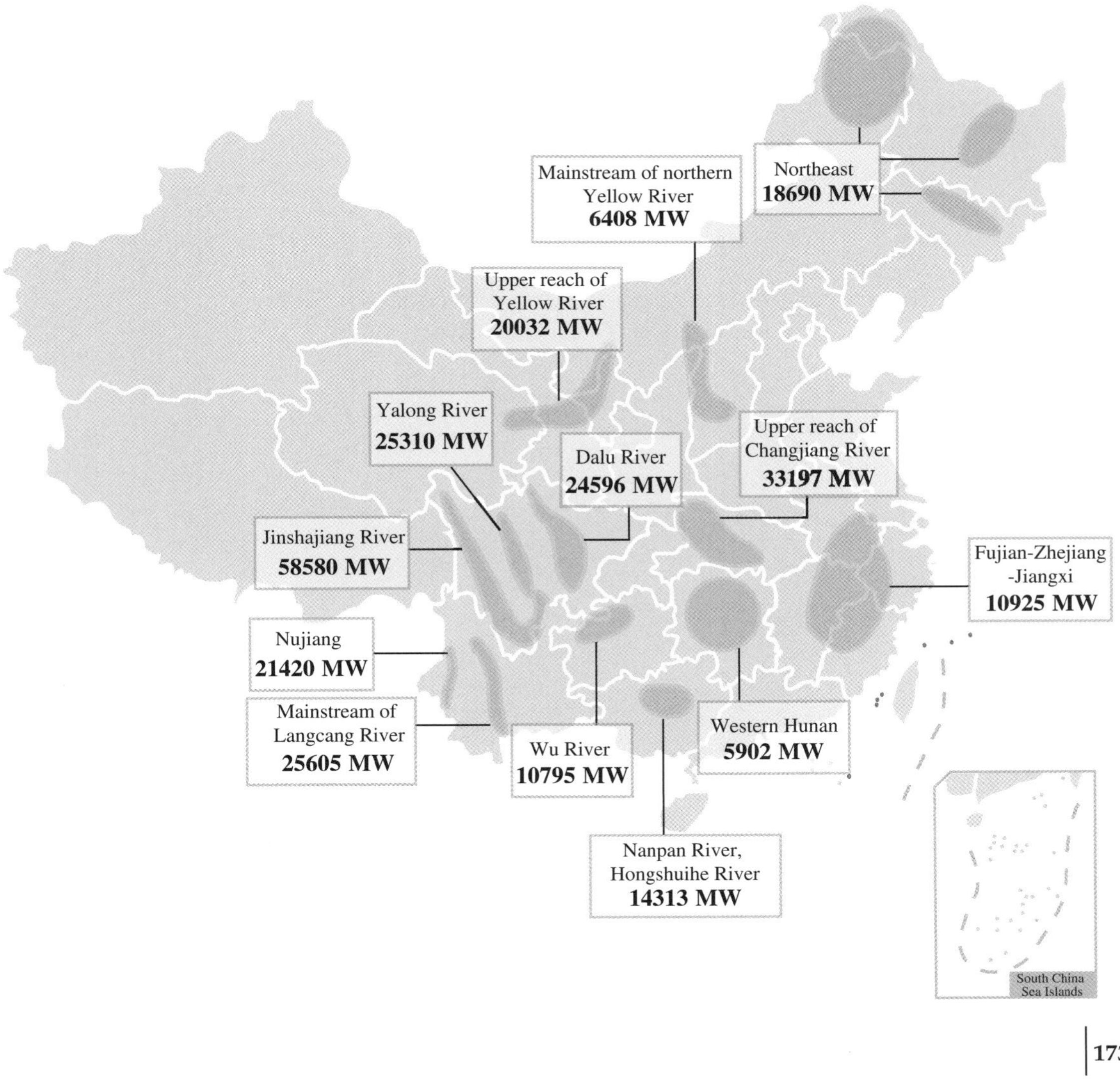

Mainstream of northern Yellow River
6408 MW
Northeast
18690 MW
Upper reach of Yellow River
20032 MW
Yalong River
25310 MW
Dalu River
24596 MW
Upper reach of Changjiang River
33197 MW
Jinshajiang River
58580 MW
Fujian-Zhejiang -Jiangxi
10925 MW
Nujiang
21420 MW
Mainstream of Langcang River
25605 MW
Wu River
10795 MW
Western Hunan
5902 MW
Nanpan River, Hongshuihe River
14313 MW
South China Sea Islands

Pattern and plan for hydropower-centered West-to-East Electricity Transmission Project

Definition →

The West-to-East electricity transmission. Project refers to developing the power resources in the provinces and regions in West China, such as Guizhou, Yunnan, Guangxi, Sichuan, Inner Mongolia, Shanxi, Shaanxi, then transmitting them to regions short of power resources like Guangdong, Shanghai, Jiangsu, Zhejiang and the Beijing-Tianjin-Tangshan area.

Strategic significance →

According to the power strategic pattern of West-to-East Electricity Transmission, North-South Power Exchange, and Nationwide Interconnection, the West-to-East Electricity Transmission Project, one of the major projects of the Grand Western Development Program was officially launched at the end of 2000. The nationwide interconnection is formed as follows: West-to-East Electricity Transmission gives rise to three power transmission thoroughfares, i.e. the south, the central and the north; North-South Power Exchange promotes the interconnection between the south, the central and the north, thus building a nationwide uniform interconnected power grid. The West-to-East Electricity Transmission Project turns the resource superiority in the western regions into an economic edge, benefiting both the western provinces and regions and the eastern ones to make good use of their specific advantages, thus promoting a balanced development of regional economies and a sustainable development of the power industry.

Achievments →

The years from 2004 to 2007 were the period when South China experienced the most severe power shortages. In this period, the capacity of West-to-East Electricity transmission increased by 1.3 times, at an annual average growth rate of 34%.

By the end of 2009, the three thoroughfares of West-to-East Electricity Transmission hit 69.98 million kW, among which that of the south thoroughfare was 24.78 million kW; that of the central thoroughfare was 17.50 million kW; and that of the north was 27.70 million kW.

By the end of 2009, among the supporting power supplies in West-to-East Electricity Transmission, the hydropower scale was 20.18 million kW, accounting for 10.3% of the nationwide hydropower installed scale; the thermal power scale was 49.80 million kW.

Three major thoroughfares of West-to-East Electricity Transmission

North thoroughfare: transmitting hydropower from the upper reach of the Yellow River and pithead thermal power from Shanxi, Shaanxi and western Inner Mongolia, to the Beijing-Tianjin-Tangshan area.

North thoroughfare

Central thoroughfare: mainly transmitting hydropower from the Three Gorges and the tributaries of the Jiangshajiang River to East China (Shanghai, Jiangsu and Zhejiang).

Central thoroughfare

South thoroughfare: transmitting the hydropower from the Wu River in Guizhou, the Lancang River in Yunnan, and the Nanpan River, Beipan River and Hongheshui River located at the juncture of the three provinces, Guizhou, Yunnan and Guangxi, and the pithead thermal power from Yunnan and Guizhou, to Guangdong Province.

South thoroughfare

Beijing-Tianjin-Tangshan

East China

Guangdong

South China Sea Islands

Planning of China's hydropower capacity in the West-to-East Electricity Transmission Project during the years 2005–2020

Unit: megawatt

Transmitting end	Export direction	2005	2010	2015	2020
Overall scale of West-to-East Electricity Transmission		**16000**	**32600**	**61200**	**83100**
South thoroughfare export		7600	11400	19300	29800
Yunnan network-to-network export	Guangdong	1600	2800	2800	2800
Langcang River point-to-network export	Guangdong			6000	9000
Middle reach of Jinshajiang River point-to-net work export	Guangdong				6000
Guizhou network-to-network export	Guangdong	1100	1600	3500	4400
Hongshuihe River export	Guangdong	1900	4000	4000	4600
Three Gorges export	Guangdong	3000	3000	3000	3000
Central thoroughfare export		8400	19700	38900	50300
Sichuan main network export	Chongqing, Central China, East China	2100	3600	3200	3100
Yalong River point-to-network export	Chongqing, Central China, East China	900	900	7400	10400
Lower reach of Jinshajiang River export	Central China, East China			13100	21600
Three Gorges export	Central China, East China	5400	15200	15200	15200
North thoroughfare export			1500	3000	3000
Hydropower of the upper reach of the Yellow River	North China		1500	3000	3000

Data source: National Energy Administration under National Development and Reform Commission, the website of the Ministry of Water Resources of the People's Republic of China at http://www.mwr.gov.cn/, the state power information network at http://www.sp.com.cn/, China energy website at http://www.china5e.com/

Nuclear power industry

2009 was the year when China's nuclear power underwent a steady rapid development. In power construction, the completed investment in nuclear power infrastructure rose by 74.9% year on year, boasting the highest investment growth rate. In 2010 investment in nuclear power infrastructure continued going up, with a completed investment of 62.9 billion yuan.

In 2009, China further accelerated the examination and approval of nuclear power projects and their construction. Within the year the projects that passed the examination were as follows: two nuclear power units each of 1.25 million kW in Sanmen, Zhejiang Province, two nuclear power units each of 1.25 million kW in Haiyang, Shandong Province, and two nuclear power units each of 1.75 million kW in Taishan, Guangdong Province, which have been put into construction successively. By the year end, nuclear power units currently under construction in China amounted to 20, with the scale of 21.92 million kW, and the status of nuclear power in the power supply structure further consolidated.

In 2010, nuclear power progressed in a steady way: a smooth headway was made in the preparatory work for the construction of inland nuclear power stations; a batch of nuclear power units were put into construction and two were put into operation; a series of key equipment realized autonomy; progress was made in the import, digestion, absorption and re-innovation of third-generation nuclear power technology from overseas; a remarkable breakthrough was made in fourth-generation nuclear power technology.

By the end of 2009, the installed capacity of China's nuclear power was 9.08 million kW, with 11 units in all, only making up 1.04% of the total power generation equipment nationwide. In 2010, China's nuclear power industry entered a scale-development phase at a rapid pace. By the year end, the nationwide nuclear power installed capacity hit 10.82 million kW, up 19.2% year on year and making up 1.1% of the total installed capacity of power generation equipment.

Installed capacity of nuclear power generation equipment in China

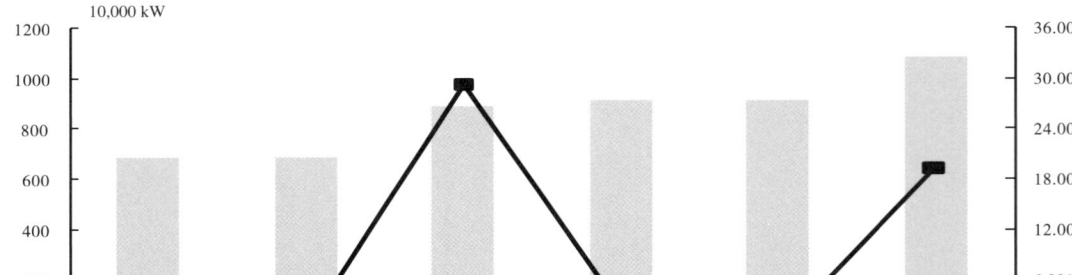

Data source: China Electricity Council

In 2009, the electricity generation of nuclear power was 66.8 billion kWh, decreasing by 2.3%, making up only 1.95% of the total electricity generation; in 2010, the nuclear power industry witnessed a rapid growth, with the year-round power generation being 73.7 billion kWh, a substantial year-on-year rise by 10.3%, making up 1.8% of the total power generation nationwide.

Nuclear power generation

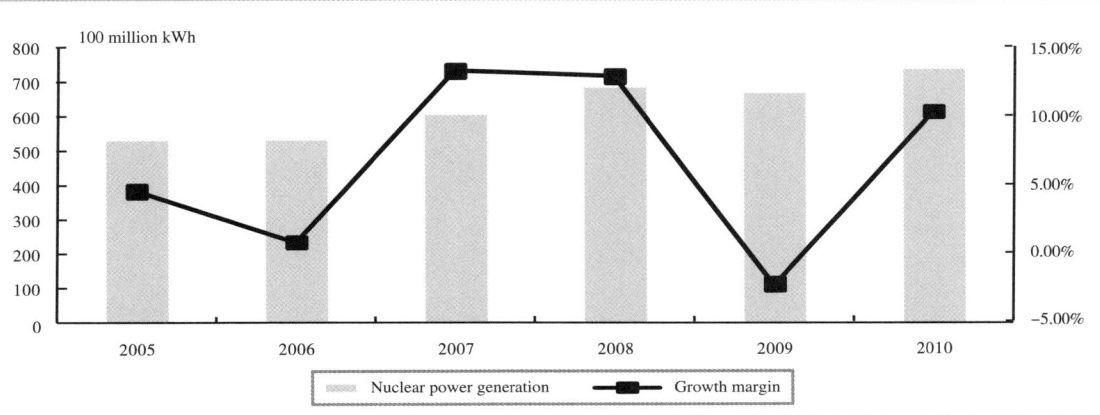

Data source: China Electricity Council

In 2009, the utilization hours of the power generation equipment in nuclear power plants were 7716 h, rising by 37 h year on year; in 2010, the utilization hours were 7924 h, up 208 h year on year.

Section 5: Typical power enterprises in China

Huaneng Power International Inc.

Company overview

Huaneng Power International Inc., China's largest listed power corporation, mainly runs the business of investing nationwide in constructing and operating large-scale thermal power-led electric power plants.

Equity structure

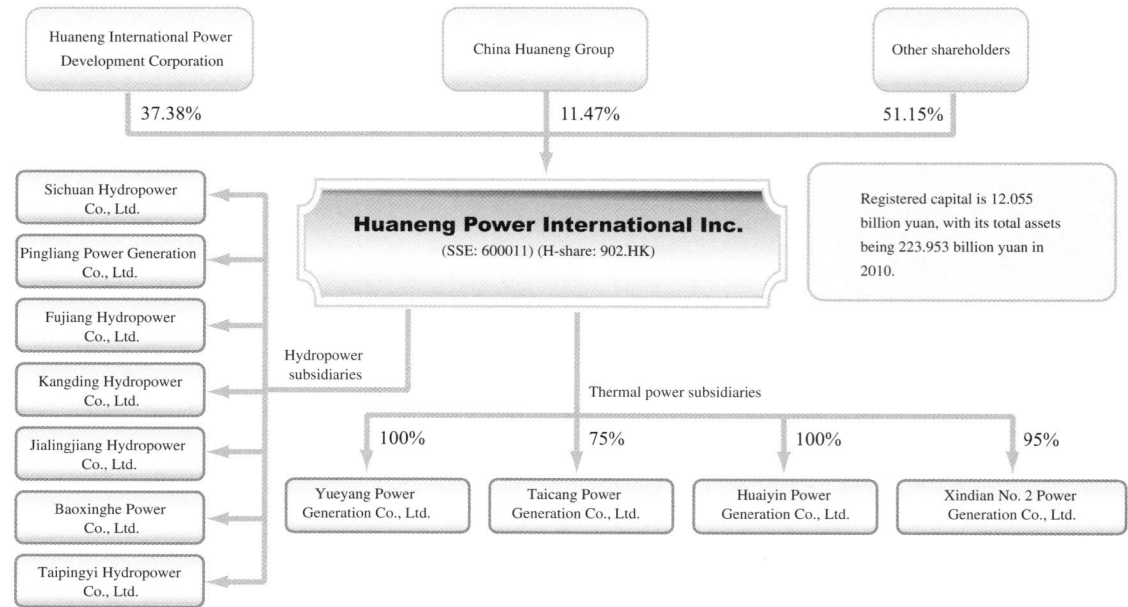

Business performance

2009

The company achieved a year-round operating revenue of 797.4 billion yuan, up 10.4% year on year; the total profit was 6.05 billion yuan, up 248.74% year on year, with a net profit of 5.08 billion yuan attributable to the shareholders of the listed companies, up 242.6% year on year. The main factor leading to the substantial growth of the net profit was the decline in steam coal prices during the reporting period.

The company's cumulative electricity generation by the electric power plants operating inside China was 203.52 billion kWh, up 10.2% year on year, the growth mainly due to the contribution made by the newly acquired power plants and the units newly put into operation.

2010

The company achieved a year-round operating revenue of 104.31 billion yuan, up 30.8% year on year, and a net profit of 3.54 billion yuan attributable to the parent company, down 30.2% year on year. The decline of its business performance was mainly caused by the increase in coal prices.

Under the effect of the units newly put into operation, in 2010 the company's combined-caliber power generation by domestic power plants rose by 26.3% (comparable-caliber power generation rose by 9.5%).

Huadian Power International Co., Ltd.

Company overview

Huadian Power International Co., Ltd. mainly runs business of constructing and operating power plants and producing and marketing power products. The power plants it currently owns all are of the thermal-power-led type, and the coverage is mainly within Shandong Province, with coal as the major production material.

Equity structure

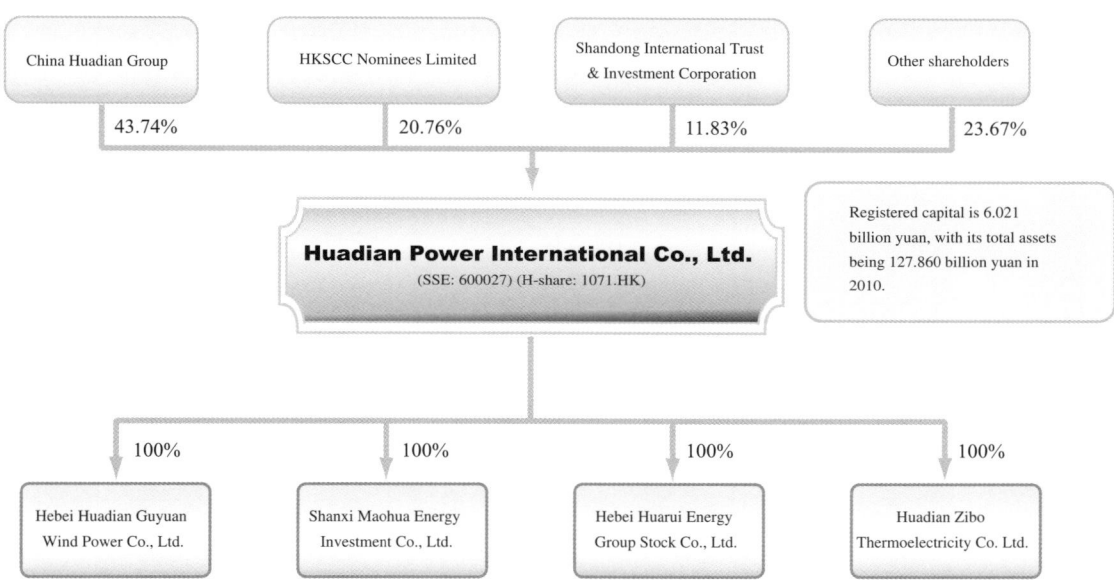

Business performance

2009

The company realized a year-round operating revenue of 36.66 billion yuan, up 14.7% year on year, achieving a net profit of 1.16 billion yuan attributable to the parent company, up 37.2% year on year. The turn from loss to gain mainly benefited from the increase in electricity generation, the decline in coal prices and the tail-raising factor in the electricity price adjustment in 2008.

The company saw a steady growth of power generation, though the utilization hours of power generation equipment slightly fell. The year-round electricity generation was 107.47 billion kWh, up 6.7% year on year, and the utilization hours of coal-fired power generation unit were 4954 h, down about 72 h year on year.

2010

The company realized a year-round operating revenue of 45.45 billion yuan, up 23.9% year on year, with a net profit of 2.1 billion yuan, down 81.9% year on year. The decline in net profit was mainly due to a substantial growth in coal prices that made the growth margin of its operating cost higher than that of its operating revenue.

The whole-year electricity generation was 1302.9 kWh, up 21.2% year on year.

GD Power Development Co., Ltd.

Company overview

GD Power Development Co., Ltd. mainly deals with the production and marketing of electric power and thermal power, power grid operation, development and application of new energy projects, hi-tech and the environmental protection industry.

Equity structure

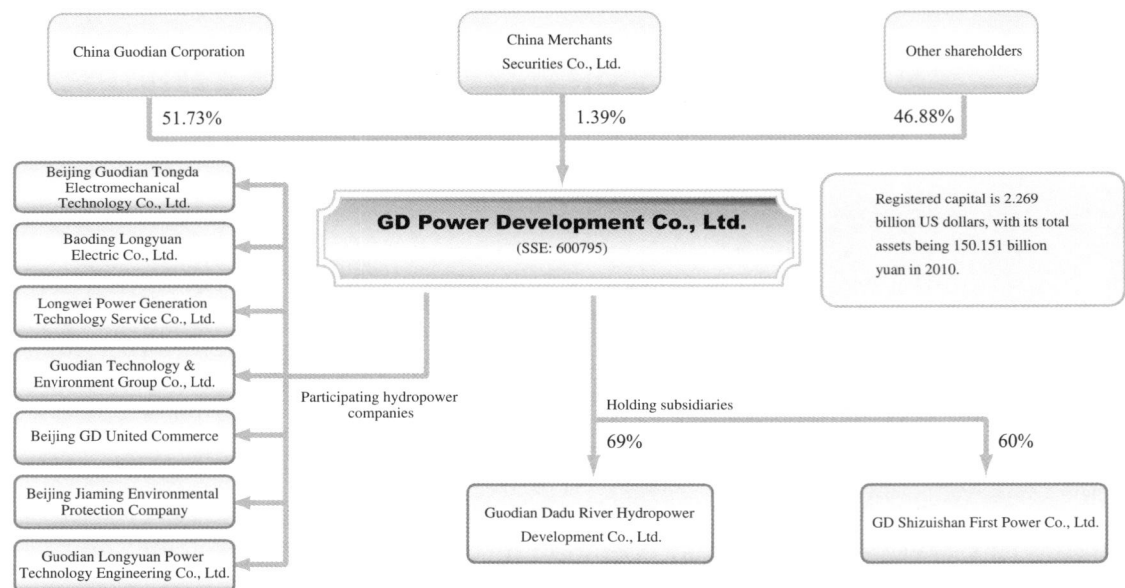

Business performance

2009

The company realized a year-round operating revenue of 19.45 billion yuan, up 18.3% year on year, the reasons being as follows: for one thing, it purchased 51% of the shares of Younglight Group and combined the chemical businesses; for another, revenues of the traditional power generation business and heating business rose respectively by 10% and 51% year on year.

The company realized a year-round net profit of 1.59 billion yuan, up 791.7% year on year, the main reason being the decline of the unit price of standard coal by 8.3%

over the previous year and the influence of the tail-raising factor in electricity price adjustment.

In 2009, the newly added installed capacity was 2.700 million kW, with electricity generation of 62.48 billion kWh, up 3.4% year on year.

2010

The company realized a year-round operating revenue of 40.77 billion yuan, up 19.9% year on year, mainly due to the increase in the electricity generation of the newly added units; it realized a net profit of 2.40 billion yuan attributable to the shareholders of the listed company, down 3.8% year on year.

Within the whole year, the cumulative electricity generation of various wholly owned and holding power generation enterprises (excluding Phase II Waigaoqiao and Younglight) amounted to 128.42 billion kWh, up 20.5% year on year.

The year-round examined wind power was 1.233 million kW; the reserved wind power resources hit 20 million kW; the examined and acquired hydropower projects hit 2.84 million kW; and that of the examined photoelectric projects amounted to 55,000 kW.

Guangdong Electric Power Development Co., Ltd.

Company overview

Guangdong Electric Power Development Co., Ltd. mainly deals in investment in, construction, operation and management of electric power projects, production and marketing of electric power, technological consultation and services for the power industry.

Equity structure

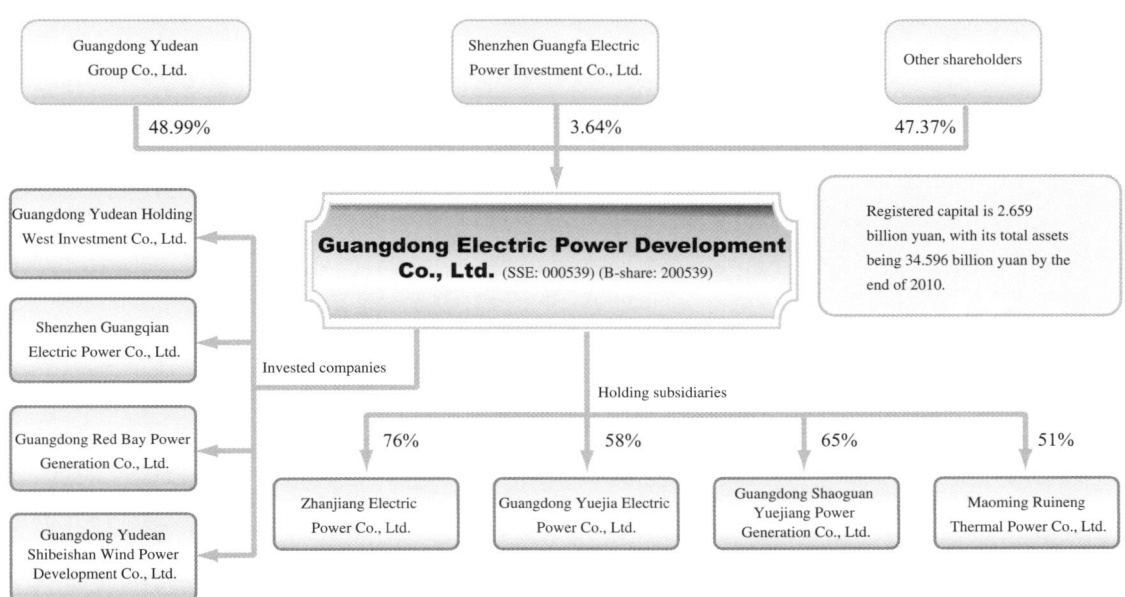

Business performance

2009

Within the whole year, the company realized a main-business operating revenue of 12.23 billion yuan, up 5.3% year on year, and a total profit of 1.66 billion yuan, with 1.16 billion yuan of net profit attributable to shareholders of the listed company, up 3921% year on year.

The large-scale growth of the year-round profits mainly benefited from the decline in coal prices and coal transport prices as well as the effective cost control. This year the fuel cost per kilowatt hour dropped by about 30% year on year.

The year-round electricity generation was 29.76 billion kWh, down 0.8% year on year, with completed on-grid energy being 27.50 billion kWh, down 0.8% year on year.

2010

The company achieved a year-round operating revenue of 12.64 billion yuan, up 3.3%, and a total profit of 1.17 billion yuan, down 29.6% year on year, with a net profit of 0.77 billion yuan attributable to shareholders of the listed company, down 34.3% year on year.

The fall in its year-round profit was mainly caused by the growth in steam coal prices.

The year-round equity power generation was 29.59 billion kWh, up 4.4% year on year, and equity on-grid energy was 27.71 billion kWh, up 5.1% year on year.

China Yangtze Power Co., Ltd.

Company overview

China Yangtze Power Co., Ltd., China's largest listed hydropower company, mainly deals with production, operation and investment in electric power, technological consultation on electric power production, and maintenance and repair of hydropower construction.

Equity structure

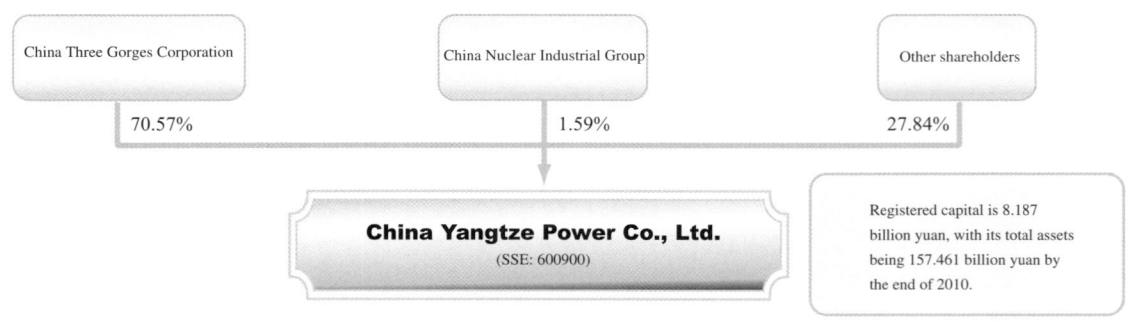

Business performance

2009

Within the whole year, the company realized a main-business operating revenue of 11.02 billion yuan, up 17.5% year on year, mainly benefiting from the smooth handing over of major assets; the total profit was 6.00 billion yuan, up 13.5% year on year, with a net profit of 4.62 billion yuan attributable to shareholders of the listed company, up 15.4% year on year.

The whole-year electricity generation of the Three Gorges hit 79.9 billion kWh, down 1.2% year on year, the margin of decline of its electricity generation beginning to narrow

2010

The company achieved a year-round operating revenue of 21.88 billion yuan, up 98.6% year on year, and a net profit of 8.23 billion yuan, up 78.1% year on year. The large growth margin of the operating revenue was mainly due to the fact that after a major asset restructuring in September 2009, 26 power-generation units of the Three Gorges Dam all belonged to the company, thus leading to a substantial increase in current electricity generation and electricity sales revenue over the previous year.

The company's Three Gorges-Gezhouba Cascade Hydroelectric Projects generated 100.6 billion kWh of electricity.

This year the company planned for the Three Gorges-Gezhouba hydro project to achieve electricity generation of 104 billion kWh, with the Three Gorges Dam and Gezhouba Dam generating 88 billion kWh and 16 billion kWh respectively.

Shenergy Company Ltd.

Company overview

Shenergy (Group) Company Ltd. mainly deals with investment in, construction, operation and management of infrastructure projects of electric power, fuel gas, etc. A wholly state-owned limited liability company, it is under the supervision of Shanghai Municipal State-owned Assets Supervision and Administration Commission which makes capital contribution to it.

Equity structure

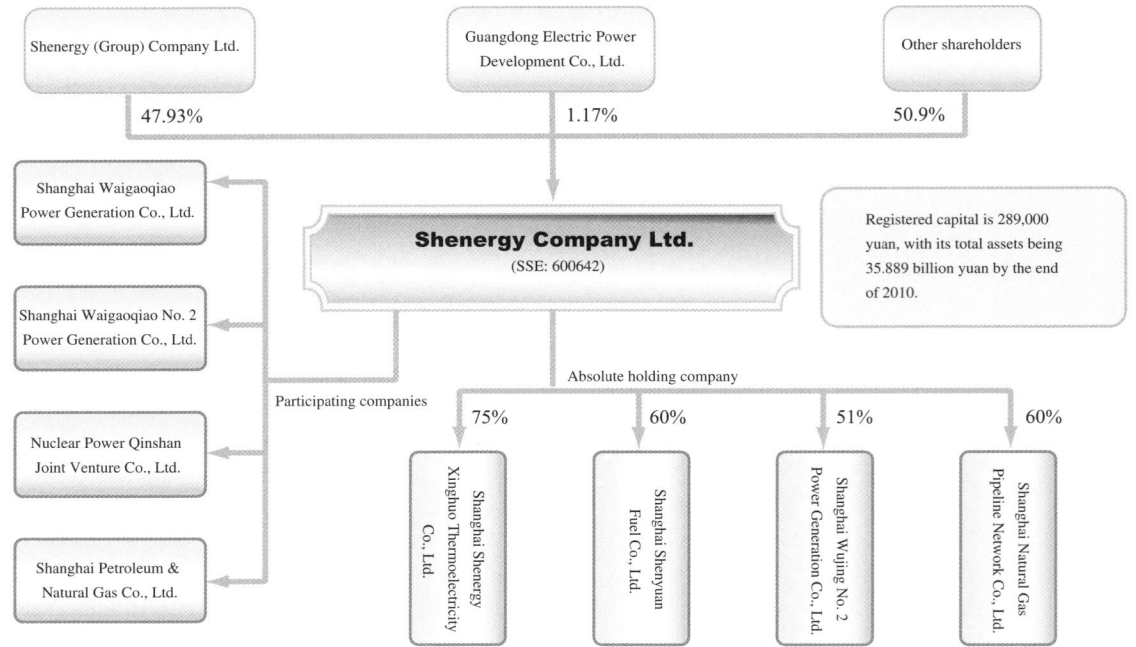

Business performance

2009

The company achieved a year-round operating revenue of 15.39 billion yuan, up 19.2% year on year, with a net profit of 1.60 billion yuan attributable to the parent company owners, up 170.1% year on year.

Power generation: In 2009 the full-caliber electricity generation was 45.9 billion kWh (excluding nuclear power), with equity electricity generation being 21.3 billion kWh. The volume of electricity maintained a year-on-year balance, but the fallback of coal prices led to a substantial profit rebound in the electric power division.

Oil and gas: Within the whole year of 2009, oil and natural gas companies exploited 91,000 tons of crude oil and produced 0.4 billion m³ of natural gas, realizing a net profit of 0.286 billion yuan, with a significant fall both in its production and profit. The natural gas pipeline network companies realized gas supply of 3.33 billion m³, achieving a net profit of 0.209 billion yuan, with a slight fall year on year.

Financial investment: In 2009 the equity profit contributed by Shenergy assets was 0.14 billion yuan, with a slight increase.

2010

The company achieved a year-round operating revenue of 19.07 billion yuan, up 23.8% year on year, realizing a net profit of 1.37 billion yuan, down 14.3% year on year. Although the whole-year power volume and operating revenue both saw a rise, yet due to the growth of coal prices its profits witnessed a decline.

Power generation: The whole-year volume of electricity grew, mainly contributed by the participating power plants. However, since the rise of coal prices caused the profit of the electric power division to fall to 0.7 billion yuan, the share of profit contribution also fell, down to 51.0%.

Oil and gas: Benefiting from the rise of oil prices and gas supply, profits saw an increase. The overall equity profit contributed about 0.27 billion yuan, up 13.0% and the share of profit contribution also went up.

Financial investment: It generated a net profit of about 0.39 billion yuan.

Profit composition of different sectors (100 million yuan)	2007	2008	2009	2010
Electric power	7.23	−0.6	10.8	7
Oil natural gas	2.3	2.6	1.15	1.37
Pipeline network companies	1.44	1.56	1.25	1.41
Shenergy assets income and earnings from other financial investment	7.6	2.37	2.8	3.94
Total	18.57	5.93	16	13.72

Datang International Power Generation Co., Ltd.

Company overview

Datang International Power Generation Co., Ltd., one of China's large-sized independent power generation companies, deals mainly with the following: construction and operation of power plants, electric power and thermal power marketing; check, repair, maintenance and debugging of power equipment; and electric power technical services. Presently, the company manages about 50 wholly owned and holding power generation companies as well as other project corporations, covering 18 provinces over the country. On March 21, 1997, H shares of Datang International Power Generation Co., Ltd. were listed in Hong Kong and London.

Equity structure

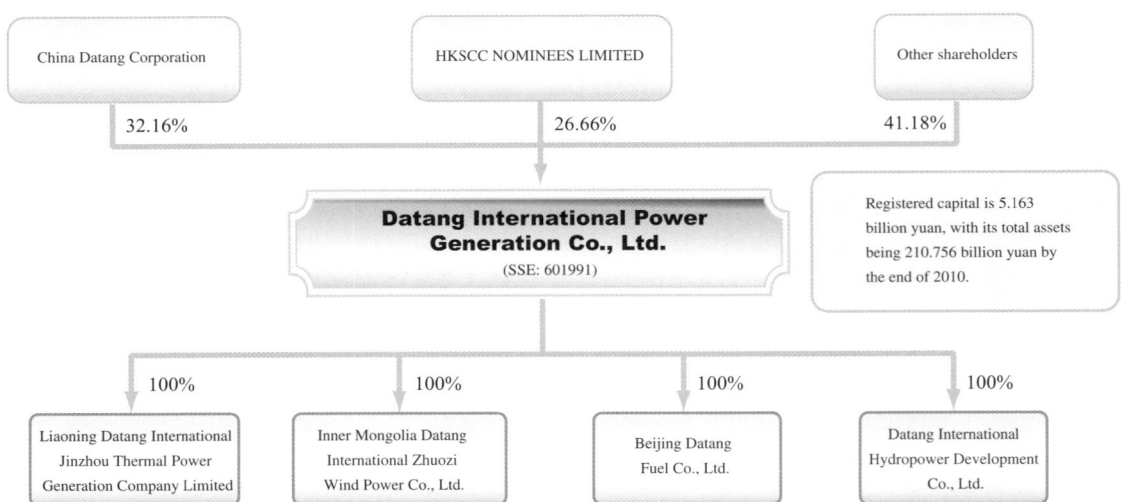

China Datang Corporation — 32.16%

HKSCC NOMINEES LIMITED — 26.66%

Other shareholders — 41.18%

Datang International Power Generation Co., Ltd.
(SSE: 601991)

Registered capital is 5.163 billion yuan, with its total assets being 210.756 billion yuan by the end of 2010.

Liaoning Datang International Jinzhou Thermal Power Generation Company Limited — 100%

Inner Mongolia Datang International Zhuozi Wind Power Co., Ltd. — 100%

Beijing Datang Fuel Co., Ltd. — 100%

Datang International Hydropower Development Co., Ltd. — 100%

Business performance

2009

Within the whole year, the company realized a main-business operating revenue of 47.94 billion yuan, up 29.9% year on year; the total profit was 3.09 billion yuan, up 538.65% year on year, with a net profit of 1.48 billion yuan attributable to shareholders of the listed company, up 131.9% year on year.

The whole-year electricity generation was 141.87 billion kWh, up 11.9% year on year, with on-grid energy being 133.55 billion kWh, up 12.1% year on year. The newly added installed capacity the year round was 5654 MW.

2010

The company achieved a year-round operating revenue of 60.67 billion yuan, up 26.6% year on year, realizing a net profit of 2.47 billion yuan attributable to the parent company, up 76.2% year on year.

The whole-year electricity generation was 178.50 billion kWh, up 26.0% year on year. The installed capacity hit 36.30 million kW, up 18.0% year on year.

New projects have been gradually put into operation, and the company's power supply structure has been further optimized. By the year end, production of thermal power hit 32.01 million kW, making up 88.0% of the total; that of hydropower was 3.86 million kW, with a 11.0% share; and that of wind power amounted to 0.43 million kW, making up 1.0% of the total.

Chapter 11   New Energy Industry

Section 1: Wind energy

China abounds in wind energy resources, especially in the coastal regions and islands in southeastern China, and northern China (i.e. Northeast China, North China and Northwest China).

According to the statistic report issued by the Global Wind Energy Council, in 2009 the world wind-power installed capacity amounted to 0.158 billion kW, with the cumulative growth rate up to 31.9%. By the year end, there were over 100 countries engaged in the development of wind power, among which 17 countries boasted a cumulative installed capacity of over 100 million kW. The world top 10 countries in terms of cumulative installed capacity were respectively the USA, China, Germany, Spain, Italy, France, Britain, Portugal and Denmark.

In 2010, wind power development worldwide slowed down, with a negative growth of 6.5% in year-round newly added installed capacity over the previous year. By the year end, worldwide cumulative installed capacity of wind power hit 0.194 billion kW, among which the Asian region saw a pretty significant growth in installed capacity, the European region maintained basically the same growth margin as that of last year, while other regions, the USA in particular, witnessed a fall. In comparison, China's newly added installed capacity of wind power accounted for 46.1% of the world total the year round, so that China surpassed the USA and became the world's top wind energy market.

In 2009, the newly added wind-power units in China (excluding Taiwan Province) hit 10,129, with the installed capacity being 13.803 million kW, up 124% year on year. It surpassed the USA as the first in the world in newly added wind-power installed capacity for the year. Its cumulative wind-power units hit 21,581, with an installed capacity of 25.805 million kW, up 114% year on year, realizing a doubled installed capacity for the fourth consecutive year.

In 2010, the newly added wind-power units in China (excluding Taiwan Province) totaled 12,904, with the installed capacity being 18.928 million kW, up 37.1% year on year; the cumulative wind-power installed units hit 34,485, with an installed capacity of 44.733 million kW, up 73.3% year on year, thus stopping the momentum of doubling.

Installed capacity of China's wind power

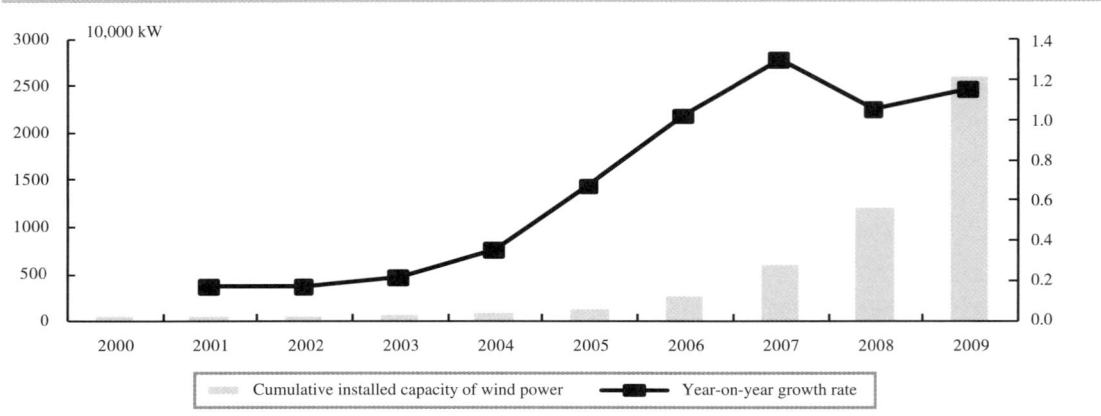

Data source: China Electricity Council

In 2009, there were in all 24 provinces and autonomous regions which built their own wind farms in China. By the year end, there were nine provinces and regions in China with a cumulative installed capacity of over million kW, among which four regions boasted a cumulative installed capacity of over 2 million kW, i.e. Inner Mongolia, Hebei, Liaoning and Jilin. Inner Mongolia Autonomous Region took the lead with its cumulative installation this year, realizing a substantial growth of 150%.

In 2010, Shaanxi, Anhui, Tianjin, Guizhou and Qinghai all realized the zero-breakthrough in wind-power installation; Beijing, Jiangxi, Guangxi and Hong Kong maintained the same installed capacity as that of the previous year; the installed capacity of Gansu Province increased rapidly. China's top 10 provinces and regions in cumulative installed capacity were respectively Inner Mongolia, Gansu, Hebei, Liaoning, Jilin, Shandong, Heilongjiang, Jiangsu, Xinjiang and Ningxia.

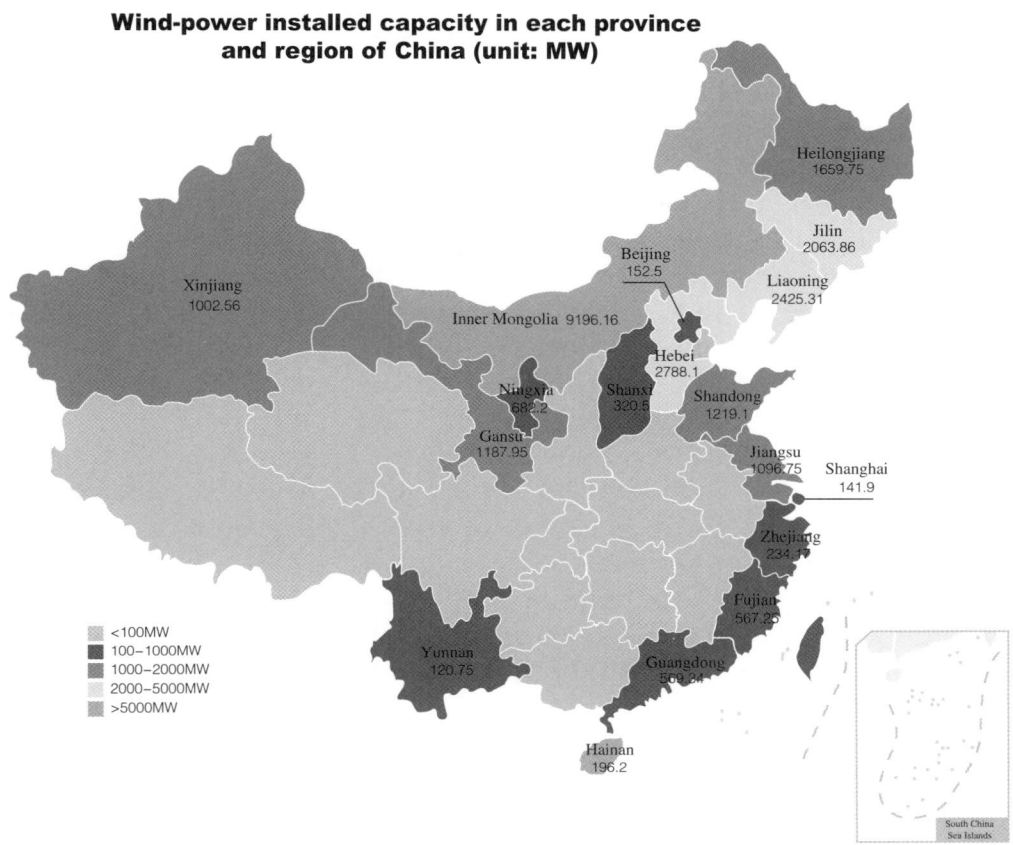

Wind-power installed capacity in each province and region of China (unit: MW)

In 2009, the nationwide wind power generation saw a significant increase of 111.1%, a doubled growth for four years running. In 2010, this amount hit 50.1 billion kWh, up 81.4% over the previous year, its growth rate higher than that of hydropower and thermal power generation.

China's wind power generation

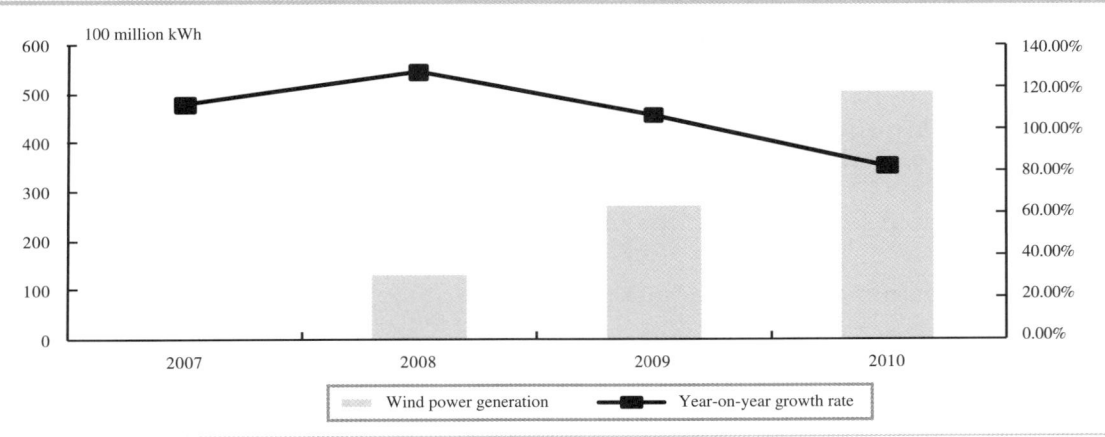

Shanghai's Donghai Bridge wind farm, our country's first, as well as Asia's first wind farm, put into operation its first batch of three units on September 4, 2009, at 21:15, formally connecting them to the power grid. This offshore wind farm consists of 34 wind power units, each with an installed capacity of 3 MW, which were manufactured by Huarui Wind Power Technology Co., Ltd. Their total installed capacity is 102 MW. Their designed annual utilization hours of wind power generation were 2624 h, with the on-grid energy being 0.267 billion kW.

China's wind power price support mechanism has gradually shifted from the price based on return on capital (ROC) to the average price achieved through the competitive bidding system in wind farm development contracts, and finally forming a fixed electricity price based on the adjustment of differentiations between wind energy resources.

On July 20, 2009, the State Development and Reform Commission (SDRC) issued the Notice Regarding the Price Policy of Wind Power Into Electricity Grid, aiming to regulate the management of wind power price and set the on-grid price of land wind power according to different resource zones.

The Notice stipulates that in terms of wind energy resources and working conditions at construction, the whole country is to be classified into four wind power resource zones to set the corresponding on-grid prices of wind power, i.e. Resource Zone I (0.51 yuan/kWh), Resource Zone II (0.54 yuan/kWh), Resource Zone III (0.58 yuan/kWh), and Resource Zone IV (0.61 yuan/kWh). From now on, any new land wind power project, including intertidal mudflats with their average tide over the high tide mark within a couple of years and island areas with permanent residents, will uniformly adopt the price set according to this standard specific to their corresponding wind energy resource zone.

During the Twelfth Five-Year period, our country will spare no efforts to elevate the share of non-fossil energy in primary energy consumption up to 11.4%, and by 2020 this share will be around 15%. In order to obtain this goal, and to complete the mission of safeguarding the energy supply and reasonably control total energy consumption, our country will make great efforts to promote highly effective utilization of clean energy and clean energy development, among which wind power development is a significant focus.

Section 2: Solar energy

Located in the Northern Hemisphere, China abounds in solar energy resources, with the nationwide annual solar radiation totaling 928–2333 kWh/m^2 and the median being 1626 kWh/m^2.

The distribution of China's annual average sunshine hours per day has the characteristic of being scarce in the southeast while rich in the northwest, and gradually increasing from southeast to northwest.

Distribution of solar energy resources in China

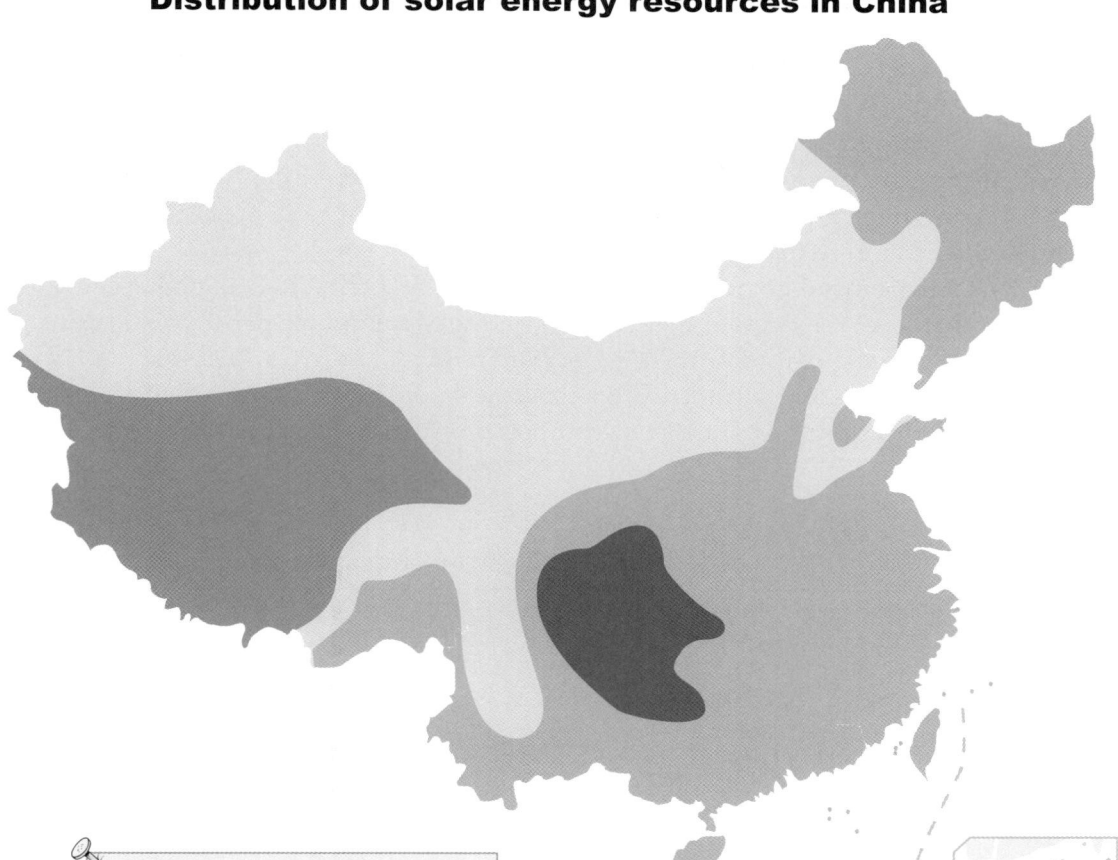

■ Rich belt   6700MJ(m²·a)
Subrich belt   5400–6700MJ/(m²·a)
General belt   4200–5400MJ/(m²·a)
■ Scarcity belt   < 4200MJ/(m²·a)

South China
Sea Islands

Distribution of the annual average sunshine hours per day in China

Regions	Annual average sunshine hours per day
North of the Huai River/Qinling Mountains line, and the plateau regions west of the eastern slopes of the Qinghai-Tibet Plateau and Yunnan Plateau	Over 2200
Qingdao-Lanzhou line, i.e. areas lying north of the latitude degrees at north, but excluding the northern and eastern parts of Northeast China	Over 2600
Inland areas northwest of the line of Xilinhaote, Yinchuan, Xining and Lhasa	Over 3000

Influenced by the financial crisis, in 2009 the global installed capacity of photovoltaic systems was around 6.6 GW, only rising by 16% over that in 2008, the lowest growth rate within the recent 10 years.

Europe is still the major solar energy market: in 2009 the capacity amounted to 5 GW, with a share of around 79%; the North American market made up 8.7%, among which the US installed capacity increased by 46% in 2009, hitting 500 MW; the Asian market made up 8.1%, among which Japan and China possessed installed capacities of 230 MW and 160 MW respectively, both with rapid growth.

In 2009, China's installed capacity rapidly went up, yet remained at a relatively lower level. Before that year, the development rate of market demand for photovoltaic power in China had always remained low. In 2008, its share in global newly added installed capacity and cumulative installed capacity was rather small, with its cumulative installed capacity making up only 1% of the world total, and its newly added installed capacity making up about 2%.

Yearly photovoltaic installed capacity in the world and China

	World total installed photovoltaic capacity (MW)	China's total installed photovoltaic capacity (MW)	China's share in the world total %
2006	1603	10	0.6%
2007	2932	20	0.7%
2008	5950	40	0.7%
2009	7380	160	2.2%
2010	16000	500	3.1%

Photovoltaic market policy

In 2009, with the carrying out of a series of favorable policies for the photovoltaic market, such as the Solar Roofs Plan, Golden Sun Demonstration Project and the Notice Regarding the Execution of the Golden Sun Demonstration Project, our country increasingly strengthened its efforts to stimulate the domestic photovoltaic market.

In April 2009, the Ministry of Finance and the Ministry of Housing and Urban-Rural Development issued first the Solar Roofs Plan: the central budget will arrange parts of the special fund for renewable energy to proactively promote the demonstration of building-integrated photovoltaics in large and medium-sized cities; it will develop in the rural and remote areas off-grid power generation; the project allowance is 20 yuan/kWh.

On July 21, 2009, the Ministry of Finance, Ministry of Science and Technology and National Energy Administration officially initiated the Golden Sun Demonstration Project. According to the Golden Sun Demonstration Project Interim Measures, the allowed capacity this time is no more than 20 MW for each province and no less than 500 MW nationwide.

On November 13, 2009, the three aforesaid ministries once again issued the Notice Regarding the Execution of the Golden Sun Demonstration Project, requiring the acceleration of the execution of the Golden Sun Demonstration Project. According to the introduction by the Ministry of Finance, the preliminary prediction indicates that the total investment in the Golden Sun Demonstration Project amounted to around 20 billion yuan, and it is planned to be completed within two to three years.

Appendix I   Brief Introduction to Energy

Brief introduction to energy

Types of energy

Energy is categorized into different types in terms of phase morphology, features or conversion and utilization. The recommended categories by the World Energy Council are as follows: solid fuel, liquid fuel, gas fuel, hydroenergy, nuclear energy, electric energy, solar energy, biomass energy, wind energy, ocean energy and geothermal energy.

Primary energy

Energy extracted from nature without any change or conversion, such as crude oil, raw coal, natural gas, biomass energy, hydroenergy, nuclear fuel, as well as solar energy, geothermal energy and tidal energy.

Secondary energy

Energy acquired through the processing or conversion of primary energy, such as coal gas, coke, gasoline, kerosene, hydropower and water hydrogen.

Conventional energy

Energy that has been produced on a large scale and widely used under the current economic and technological conditions, such as coal, oil, natural gas, hydroenergy and nuclear fission energy. Conventional energy is so called relative to new energy, the kind of energy that aims to be systematically developed and utilized based on new technology, such as solar energy, ocean energy, geothermal energy and biomass energy. The majority of types of new energy are natural and renewable, the foundation of a global lasting energy system in the future.

Commercial energy

As the kind of energy consumed in a large amount in the commodity circulation, it currently mainly includes five types, i.e. coal, oil, natural gas, hydropower and nuclear power.

Non-commercial energy

Energy utilized in situ, such as firewood, agricultural waste, and is usually renewable.

Environmental energy

Energy such as energy flow and solar energy reserved in the Earth's environment, radioactive sources within the Earth and the motion of the solar system, the initial energy of all environmental energy in the world.

Appendix II Unit Conversions

Unit conversions relevant to crude oil*

	Ton	Kiloliter	Barrel	US gallon	Ton/year
1 ton	1.000	1.165	7.330	307.860	—
1 kiloliter	0.8581	1.000	6.2898	264.170	—
1 barrel	0.1364	0.159	1.000	42.000	—
1 US gallon	0.00325	0.0038	0.0238	1.000	—
1 barrel/day	—	—	—	—	49.800

*Based on average global gravity.

Unit conversions relevant to oil products**

Products	Barrel to ton	Ton to barrel	Kiloliter to ton	Ton to kiloliter
LPG	0.086	11.600	0.542	1.844
Gasoline	0.118	8.500	0.740	1.351
Kerosene	0.128	7.800	0.806	1.240
Diesel	0.133	7.500	0.839	1.192
Residual fuel oil	0.149	6.700	0.939	1.065

**LPG is short for liquefied petroleum gas.

Unit conversions relevant to natural gas and LNG***

	Billion cubic meters NG	Billion cubic feet NG	Million toe	Million tons LNG	Million British thermal units (BTU)	Million toe
Billion cubic meters NG	1.000	35.300	0.900	0.730	36.000	6.290
Billion cubic feet NG	0.028	1.000	0.026	0.021	1.030	0.180
Million toe	1.111	39.200	1.000	0.805	40.400	7.330
Million tons LNG	1.380	48.700	1.230	1.000	52.000	8.680
Million British thermal unit (BTU)	0.028	0.980	0.025	0.020	1.000	0.170
Million toe	0.160	5.610	0.140	0.120	5.800	1.000

***NG is short for natural gas and LNG for liquefied natural gas.

1 ton of oil is approximately equivalent to

Thermal unit	10 million kilocalories
	42 gigajoules
	40 million British thermal units
Solid fuels	1.5 tons of hard coal
	3 tons of lignite
Gas fuels	See also the table natural gas and LNG
Electricity	12 megawatt hours

1 million tons of oil products is equivalent to 4500 giga-watt hour (GWH) ($4.5 terawatt hour) in electricity.

Other unit conversions:

1 ton = 2204.62 pounds = 1.1023 short tons
1 kiloliter = 6.2898 barrels
1 kiloliter = 1 cubic meter
1 kilocalories = 4.187 kilojoules = 3.968 British thermal units
1 kilojoule = 0.239 kilocalories = 0.948 British thermal units
1 British thermal units = 0.252 kilocalories = 1.055 kilojoules
1 kilocalories/hour = 860 kilocalories = 3600 kilojoules = 3412 British thermal units

Appendix III   Regional Coverage

Region or Organization	Countries and regions covered
North America	USA, including Puerto Rico, Canada and Mexico
Latin America	Caribbean (including Puerto Rico), Central America and South America
Europe	EU members, and Albania, Bosnia and Herzegovina, Belgium, Croatia, Cyprus, Macedonia, Gibraltar, Malta, Serbia, Montenegro and Slovenia
Former Soviet Union	Armenia, Azerbaijan, Belarus, Estonia, Georgia, Kazakhstan, Kyrgyzstan, Latvia, Lithuania, Moldova, Russia, Tajikistan, Ukraine and Uzbekistan
Europe and Eurasia	Including countries within Europe and the former Soviet Union
Middle East	Arabian Peninsula, Iraq, Iran, Jordan, Israel and Syria
North Africa	Including the area from Northern Egypt to Western Sahara
West Africa	From Mauritania to Angola, including Cape Verde and Lake Chad
East Africa and South Africa	From Sudan to South Africa, also including Botswana, Madagascar, Malawi, Namibia, Uganda, Zambia and Zimbabwe
Asia-Pacific Region	Brunei, Cambodia, Mainland China, Hong Kong Special Administrative Region of China, Indonesia, Japan, Laos, Malaysia, Mongolia, North Korea, the Philippines, Singapore, South Asia (Afghanistan, Bangladesh India, Burma, Nepal, Pakistan, Sri Lanka), South Korea, Taiwan China, Thailand, Vietnam, Australia, New Zealand, Papua New Guinea
Oceania	Australia, New Zealand
OECD	Austria, Belgium, Czech Republic, Denmark, Estonia, Finland, France, Germany, Greece, Congo, Republic of Ireland, Italy, Lithuania, Luxembourg, Malta, the Netherlands, Poland, Puerto Rico, Slovakia, Slovenia, Spain, Sweden, Britain, Australia, Canada, Mexico, New Zealand, Korea, the USA
OPEC members	Middle East: Iran, Iraq, Kuwait, Qatar, Saudi Arabia, United Arab Emirates (UAE) North Africa: Algeria, Libya West Africa: Nigeria Asia-Pacific Region: Indonesia South America: Venezuela
EU members	Austria, Belgium, Bulgaria, Cyprus, Czech Republic, Denmark, Estonia, Finland, France, Germany, Greece, Hungary, Republic of Ireland, Italy, Latvia, Lithuania, Luxembourg, Malta, the Netherlands, Poland, Portugal, Romania, Slovakia, Slovenia, Spain, Sweden, Britain

Appendix IV China's Top 100 Coal Enterprises in 2009

Ranking	Organization	Operating revenue (10,000 yuan)
1	Shenhua Group Corporation Limited	14402329
2	Shanxi Coal Transportation and Sales Group Co., Ltd.	9023940
3	Henan Coal Chemical Industry Group Co., Ltd.	8211530
4	Shanxi Coking Coal Group International Trading Co., Ltd.	7179795
5	China National Coal Group Corporation	7125265
6	China Pingmei Shenma Energy & Chemical Group Co., Ltd.	6813417
7	Shanxi Datong Coal Mine Group Co., Ltd.	5132418
8	Shanggong Yankuang Group Co., Ltd.	4679232
9	Shanxi Jincheng Anthracite Mining Group Co., Ltd	4455018
10	Jizhong Energy Group Co., Ltd.	4100284
11	Shanxi Lu'an Mining Group Co., Ltd.	3524840
12	Heilongjiang Longmei Mining Group Co., Ltd.	2474636
13	Shanxi Yangquan Coal Industry (Group) Co., Ltd.	3436799
14	Hebei Kailuan (Group) Co., Ltd.	3344674
15	Shandong Xinwen Mining Group Co., Ltd.	3287262
16	Shandong Zaozhuang Coal Mining Group Co., Ltd.	2962848
17	Anhui Huainan Mining Group Co., Ltd.	2767789
18	Shaanxi Coal and Chemical Industry Group Co., Ltd.	2340783
19	Anhui Huaibei Mining Group Co., Ltd.	2298639
20	Inner Mongolia Yitai Group Co., Ltd.	2291809
21	Shanxi Coal Import and Export Group Co., Ltd.	2190292
22	Jiangsu Xuzhou Coal Mining Group Corp.	1856560
23	CPI Mengdong Energy Group Co., Ltd.	1731471
24	Henan Shenhuo Group Co., Ltd.	1606158
25	China Coal Technology & Engineering Group Corp.	1490234
26	Shandong Zibo Mining Group Co., Ltd.	1363347
27	Henan Zhengzhou Coal Industry (Group) Co., Ltd.	1356088
28	Chongqing Energy Investment Group	1237827
29	Anhui Wanbei Coal And Power Group Co., Ltd.	1212764
30	Liaoning Tiefa Coal Industry (Group) Co., Ltd.	1209754
31	Henan Yima coal Industry (Group) Co., Ltd.	1189760

(Continued)

200

(Continued)

Ranking	Organization	Operating revenue (10,000 yuan)
32	Beijing Jingmei Group Co., Ltd.	930584
33	Fujian Coal Industry (Group) Co., Ltd.	898834
34	Liaoning Shenyang Coal Industry (Group) Co., Ltd.	859260
35	Shandong Feicheng Mining Group Co., Ltd.	839984
36	Shanxi Lanhua Coal Industry Group Co., Ltd.	763260
37	Liaoning Fuxin Mining Group Co., Ltd.	747984
38	Sichuan Coal Industry Group Co., Ltd.	731636
39	Inner Mongolia Yidong Coal Group Co., Ltd.	727485
40	Inner Mongolia Huineng Coal & Power Group	704877
41	Jiangxi Coal group Corporation	703342
42	Shandong Linyi Mining Group Co., Ltd.	626790
43	Liaoning Fushun Mining Group Co., Ltd.	624343
44	Yunnan Dongyuan Coal Industry Group Co., Ltd.	591684
45	China Coal Mine Construction Group Corporation	577210
46	Guizhou Panjiang Coal (Group) Co., Ltd.	553221
47	Shandong Longkou Mining Group Co., Ltd.	548167
48	Gansu Huating Coal Industry Group Co., Ltd.	546375
49	SDIC Xinji Energy Stock Co., Ltd.	542626
50	Shendong Tianlong Group Co., Ltd.	533731
51	Huaneng Hulun Buir Energy Development Co., Ltd.	523453
52	Hunan Coal Industry Group	508808
53	Inner Mongolia Mengtai Coal Electricity Co., Ltd.	498035
54	Guizhou Shuicheng Coal Mining (Group) Co., Ltd.	471822
55	Shanxi Qinxin Coal and Coke Corp.	458931
56	Shanxi Golden Ocean Energy Group Co., Ltd.	451700
57	Inner Mongolia Pingzhuang Coal Industry Group Co., Ltd.	451324
58	Shandong Taifeng Mine Industry Group Co., Ltd.	407802
59	Inner Mongolia Ximeng Technology Industry and Trade Group Co., Ltd.	398337
60	Inner Mongolia Manshi Coal Group Co., Ltd.	391421
61	Henan Zhengzhou Coal Mining Machinery Group Co., Ltd.	372234
62	Inner Mongolia Qinghua Group Co., Ltd.	357794
63	Gansu Jingyuan Coal Industry Group Co., Ltd.	333072
64	Qinghai Qinghua Mining, Smelting and Coking Group Co., Ltd.	320354
65	Shanxi Xiangyuan Local State-owned Xiangyuan Coal Mine	316008
66	Shaanxi Binxian Coal Co., Ltd.	309286

(Continued)

(Continued)

Ranking	Organization	Operating revenue (10,000 yuan)
67	Inner Mongolia MF Coal Co., Ltd.	299452
68	Shanxi Fenhe Coke and Coal Share Holding Co., Ltd.	279205
69	Hebei Cixian Liuhe Industry Co., Ltd.	275219
70	Jilin Liaoyuan Coal Mining Group Co., Ltd.	271278
71	Shandong Jining Coal Industry Group	257542
72	Shandong Fengyuan Coal Industry & Electric Power Co., Ltd.	237416
73	Jilin Tonghua Mining (Group) Co., Ltd.	201501
74	Shandong Yulong Mining Industry Group Co., Ltd.	171852
75	Gansu Yaojie Coal & Electricity Co., Ltd.	170474
76	Yunnan Xiaolongtan Mining Administration	170077
77	Shanxi Sanyuan Coal Industrial Co., Ltd.	166283
78	SDIC Coal Zhengzhou Energy Development Co., Ltd.	157908
79	Sany Heavy Equipment Co., Ltd.	157764
80	Guizhou Liuzhi Industrial and Mining Group Co., Ltd.	150258
81	Inner Mongolia Tehong Coal and Electricity Group Co., Ltd.	150012
82	Shanxi Liliu Coke & Coal Group Co., Ltd.	149339
83	Pingdingshan Coal Mine Machinery Co., Ltd.	137989
84	Shandong Honghe Mining Industry Group Co., Ltd.	132854
85	Shanxi Pingyang Heavy Industry Machinery Co., Ltd.	124735
86	Jiangsu Hong'an Group Co., Ltd.	116543
87	Shandong Mining Machinery Group Co., Ltd.	114671
88	Xinjiang Coking Coal (Group) Co., Ltd.	107804
89	Shandong Wangchao Coal Electric Power Group Co., Ltd.	105438
90	Zhejiang Changguang Group Co., Ltd.	97635
91	Hebei Cixian Shenjiazhuang Coal Mine	97181
92	Shaanxi Yulin Yushen Coal Co., Ltd.	86610
93	Shanxi Yangquan Yinying Coal Mine	86500
94	Liaoning Beipiao Coal Industry Co., Ltd.	85986
95	Shanxi Datong Que'ershan Clean Coal Co., Ltd.	85961
96	Zhengzhou Dengcao Group Co., Ltd.	85359
97	Shanxi Yitang Coal Industry Co., Ltd.	82098
98	Shandong Daizhuang Shengjian Coal Mine	78412
99	Shanxi Changzhi Jingfang Coal Co., Ltd.	73355
100	Shanxi Taiyuan Dongshan Coal Mine Co., Ltd.	66547

Appendix V   China's Top 100 Coal Enterprises in 2010

Ranking	Organization	Operating revenues (10,000 yuan)
1	Shenhua Group Corporation Limited	16124950
2	Henan Coal Chemical Industry Group Co., Ltd.	10409527
3	China Pingmei Shenma Energy & Chemical Group Co., Ltd.	8016013
4	Shanxi Coking Coal Group the International Trading Co., Ltd.	7747769
5	Shanxi Coal Transportation and Sales Group Co., Ltd.	7243878
6	China National Coal Group Corporation	7017192
7	Jizhong Energy Group Co., Ltd.	5808577
8	Hebei Kailuan (Group) Co., Ltd.	5593860
9	Shanxi Jincheng Anthracite Mining Group Co., Ltd	5543456
10	Shanggong Yankuang Group Co., Ltd.	5261887
11	Shanxi Lu'an Mining Group Co., Ltd.	4985778
12	Shanxi Yangquan Coal Industry (Group) Co., Ltd.	4960041
13	Shanxi Datong Coal Mine Group Co., Ltd.	4254301
14	Shandong Xinwen Mining Group Co., Ltd.	3683000
15	Anhui Huainan Mining Group Co., Ltd.	3524321
16	Heilongjiang Longmei Mining Group Co., Ltd.	3261532
17	Shaanxi Coal and Chemical Industry Group Co., Ltd.	3208783
18	Shandong Zaozhuang Coal Mining Group Co., Ltd.	2980688
19	Inner Mongolia Yitai Group Co., Ltd.	2589482
20	Anhui Huaibei Mining Group Co., Ltd.	2468663
21	Shanxi Coal Import and Export Group	2302888
22	Jiangsu Xuzhou Coal Mining Group Corp.	1943083
23	China Coal Technology & Engineering Group Corp.	1760412
24	Henan Zhengzhou Coal Industry (Group) Co., Ltd.	1581097
25	Henan Shenhuo Group Co., Ltd.	1480684
26	Henan Yima coal Industry (Group) Co., Ltd.	1416276
27	Shandong Zibo Mining Group Co., Ltd.	1398781
28	Anhui Wanbei Coal And Power Group Co., Ltd.	1351990
29	Chongqing Energy Investment Group	1297985
30	Liaoning Tiefa Coal Industry (Group) Co., Ltd.	1287688
31	Fujian Energy Group Co., Ltd.	1109839

(Continued)

(Continued)

Ranking	Organization	Operating revenues (10,000 yuan)
32	Liaoning Shenyang Coal Industry (Group) Co., Ltd.	1021306
33	Inner Mongolia Huineng Coal & Power Group	976541
34	Shanxi Lanhua Coal Industry Group Co., Ltd.	904462
35	Shandong Linyi Mining Group Co., Ltd.	809093
36	Beijing Jingmei Group Co., Ltd.	802130
37	Liaoning Fuxin Mining Group Co., Ltd.	781216
38	Inner Mongolia Yidong Coal Group Co., Ltd.	779427
39	Inner Mongolia Mengtai Coal Electricity Co., Ltd.	778694
40	Sichuan Coal Industry Group Co., Ltd.	767022
41	China Coal Mine Construction Group Corporation	720431
42	Shandong Longkou Mining Group Co., Ltd.	702744
43	Jiangxi Coal Group Corporation	701812
44	Shandong Feicheng Mining Group Co., Ltd.	694303
45	Jilin Coal Industry Group Co., Ltd.	672976
46	Hunan Coal Industry Group	610389
47	Liaoning Fushun Mining Group Co., Ltd.	598814
48	Gansu Huating Coal Industry Group Co., Ltd.	573224
49	Guizhou Panjiang Coal (Group) Co., Ltd.	571271
50	Yunnan Dongyuan Coal Industry Group Co., Ltd.	569804
51	Huaneng Hulun Buir Energy Development Co., Ltd.	564739
52	Henan Zhengzhou Coal Mining Machinery Group Co., Ltd.	516028
53	Inner Mongolia Manshi Coal Group Co., Ltd.	515935
54	Guizhou Shuicheng Coal Mining (Group) Co., Ltd.	502858
55	Shendong Tianlong Group Co., Ltd.	501490
56	SDIC Xinji Energy Stock Co., Ltd.	494988
57	Huolinhe Opencut Coal Industry Corp. Ltd. of Inner Mongolia	478944
58	Inner Mongolia Pingzhuang Coal Industry Group Co., Ltd.	470134
59	Shandong Taifeng Mine Industry Group Co., Ltd.	463193
60	Shaanxi Binxian Coal Co., Ltd.	420928
61	Inner Mongolia Ximeng Group Corporation	407499
62	Gansu Jingyuan Coal Industry Group Co., Ltd.	367864
63	Qinghai Qinghua Mining, Smelting and Coking Group Co., Ltd.	341152
64	Shanxi Qinxin Energy Group Corporation	329700
65	Inner Mongolia Taixi Coal Group Co., Ltd.	300841
66	Shanxi Sanyuan Coal Industrial Co., Ltd.	266985
67	Ordos Wulan Coal Group Co., Ltd.	257397

(Continued)

(Continued)

Ranking	Organization	Operating revenues (10,000 yuan)
68	Shandong Jining Coal Industry Group	249733
69	Inner Mongolia Tehong Coal and Electricity Group Co., Ltd.	227040
70	Hebei Cixian Liuhe Industry Co., Ltd.	218027
71	Sany Heavy Equipment Co., Ltd.	197413
72	Shandong Fengyuan Coal Industry & Electric Power Co., Ltd.	197365
73	Shandong Yulong Mining Industry Group Co., Ltd.	185873
74	Shanxi Xiangning Coking Coal Group Co., Ltd.	179582
75	Yunnan Xiaolongtan Mining Administration	174938
76	Guizhou Liuzhi Industrial and Mining Group Co., Ltd.	169164
77	Gansu Yaojie Coal & Electricity Co., Ltd.	165801
78	Shanxi Fenhe Coke and Coal Share Holding Co., Ltd.	164003
79	Shanxi Changzhi Jingfang Coal Industry Co., Ltd.	156866
80	SDIC Henan New Energy Development Corporation	155802
81	Pingdingshan Coal Mine Machinery Co., Ltd.	153353
82	Shaanxi Yulin Yushen Coal Co., Ltd.	152713
83	Shandong Honghe Mining Industry Group Co., Ltd.	148469
84	Shanxi Pingyang Heavy Industry Machinery Co., Ltd.	146499
85	Hebei Cixian Shenjiazhuang Coal Mine	136815
86	Sichuan Xinfu Coal Industy Co., Ltd.	129394
87	Shandong Tiansheng Coal Mine Equipment Co., Ltd.	125968
88	Shanxi Liliu Coke & Coal Group Co., Ltd.	125864
89	Shandong Mining Machinery Group Co., Ltd.	105624
90	Zhengzhou Dengcao Group Co., Ltd.	105246
91	Inner Mongolia MF Coal Co., Ltd.	97305
92	Shandong Daizhuang Shengjian Coal Mine	95032
93	Zhejiang Changguang Group Co., Ltd.	92234
94	Shanxi Yangquan Yinying Coal Mine	90238
95	Shandong Huaning Coal Mining Group Corporation Ltd.	88611
96	Shanxi Yitang Coal Industry Co., Ltd.	82794
97	Liaoning Beipiao Coal Industry Co., Ltd.	73210
98	Xinjiang Coking Coal (Group) Co., Ltd.	73140
99	Shandong Qiwu Shengjian Coal Mine	67526
100	Shanxi Taiyuan Dongshan Coal Mine Co., Ltd.	62092